CHEMICAL ENGINEERING METHODS AND TECHNOLOGY

CARBON DIOXIDE CAPTURE AND SEQUESTRATION

AN INTEGRATED OVERVIEW OF AVAILABLE TECHNOLOGIES

CHEMICAL ENGINEERING METHODS AND TECHNOLOGY

Additional books in this series can be found on Nova's website under the Series tab.

Additional e-books in this series can be found on Nova's website under the e-book tab.

ENVIRONMENTAL SCIENCE, ENGINEERING AND TECHNOLOGY

Additional books in this series can be found on Nova's website under the Series tab.

Additional e-books in this series can be found on Nova's website under the e-book tab.

CHEMICAL ENGINEERING METHODS AND TECHNOLOGY

CARBON DIOXIDE CAPTURE AND SEQUESTRATION

AN INTEGRATED OVERVIEW OF AVAILABLE TECHNOLOGIES

JOÃO FERNANDO PEREIRA GOMES

New York

For permission to use material from this book please contact us:
Telephone 631-231-7269; Fax 631-231-8175
Web Site: http://www.novapublishers.com

Additional color graphics may be available in the e-book version of this book.

Library of Congress Cataloging-in-Publication Data

Carbon dioxide capture and sequestration:an integrated overview of available technologies / Joao Fernando Pereira Gomes.
pages cm.
Includes index.
ISBN 978-1-62257-187-1 (soft cover)
1. Carbon dioxide mitigation. 2. Carbon dioxide--Absorption and adsorption. I. Gomes, Joao Fernando Pereira.
TD885.5.C3C365 2013
628.5'32--dc23

2012035206

Published by Nova Science Publishers, Inc. † New York

CONTENTS

PREFACE

This publication is intended as a university manual describing both the problem of CO_2 emissions and also the actual available sequestration technologies and measures that can be undertaken to solve the problem. It includes a comprehensive description of the actual state-of the-art, including recent scientific developments and is mainly intended for scientists and students. It is based on relevant reviews on part of the issues on this theme, as follows:

Figueroa, J., *et al.*, International Journal of Greenhouse Gas Control, 2, 9-20 (2008).
Yang, H., *et al.*, Journal of Environmental Studies, 20, 14-27 (2008).
Olajire, A., Energy, 35, 2610-2628 (2010).

CO_2 capture from gaseous effluents is, nowadays, one of the great challenges faced by chemical and environmental engineering, as the increase of CO_2 levels in the Earth atmosphere is endangering the support of living species in this planet and might also be responsible for dramatic climate changes. Several capture technologies actually exist although the only proven and mature technology is, currently, chemical absorption using aqueous amine solutions, which still poses several technological and operational problems which as not yet overcome completely. Other concurrent technologies such as adsorption and membrane are, yet, in a less developed stage and a considerable research effort has still to be done in order to have those as viable industrial processes. Other emerging technologies such as the use of ionic liquids and biological processes also need further development. Nevertheless, several alternatives exist for CO_2 reutilization as this is an abundant carbon source that could be easily used as a raw material for other carbon based raw materials, fuels and polymers.

ABBREVIATIONS

ABC	Ammonium bicarbonate
AC	Ammonium carbonate
BPL	British Physical Laboratories
BTU	Energy consumption
CANMET	Canada Center for Mineral and Energy
CA	Carbon anhydrase
CAP	Chilled Ammonia Process
CCS	Carbon Capture and Sequestration
CLC	Chemical Looping Combustion
CVD	Chemical Vapor Deposition
DEA	Diethanolamine
DOE	Department of Energy (USA)
ECBM	Enhanced Coal bed Methane
EDA	Ethylenediamine
EOR	Enhanced Oil Recovery
FTM	Facilitated Transport Membrane
GDP	Gross Domestic Product
GHG	Green House Gases
HAD	1,6-Hexanediamine
HFCLM	Hollow Fiber Contained Liquid Membrane
IEA	International Energy Agency
IGCC	Integrated Gasification Combined Cycle
IL	Ionic Liquid
ILM	Immobilized Liquid Membrane
IPCC	Intergovernmental Panel on Climate Change
ITC	International Test Centre
MCM	Mobil Crystalline Material

MDEA	Methyl-diethanolamine
MEA	Monoethanolamine
MHI	Mitsubishi Heavy Industries
MOF	Metal Organic Frameworks
NETL	National Energy Technology Laboratory
ppb	part per billion
PEI	Polyethyleneimine
POP	Population
ppm	part per million
ppt	part per trillion
PZ	Piperazine
RTIL	Room Temperature Ionic Liquid
SBA	Mesoporous Silica
TEPA	Tetraethylenepentamine
toe	ton of oil equivalent
ton C eq	ton of carbon equivalent
TSIL	Task Specific Ionic Liquid
USA	United States of America
v	Volume
VHTS	Virtual High Throughput Screening
ZIF	Zinc Imidazolate Frameworks

Chapter 1

INTRODUCTION: PROBLEM SITUATION ON CARBON DIOXIDE EMISSIONS

"Whosoever shall be found guilty of burning coal shall suffer the loss of his head"
King Edward I of England, circa 1300

The problem of carbon dioxide emissions is part of the general problem of air pollution, which has been long ago recognized, mainly in what concerns the protection of public health due to the presence of high toxicity airborne pollutants. The historic citation mentioned above, which is particularly concerned with coal combustion for generating energy, shows that this problem has been quite sensible for populations and governments throughout ages.

Carbon dioxide emissions result from the combustion reaction of carbon in a highly oxidizing atmosphere, containing 21% of oxygen, such as the terrestrial atmosphere, as follows:

$C + O_2 \rightarrow CO_2$ Reaction 1: Complete combustion
$H_2 + \frac{1}{2} O_2 \rightarrow H_2O$ Reaction 2: Water vapor formation

As those are exothermic reactions, combustion is currently used for producing energy, in thermal power plants and also in industrial processes and for fuelling internal combustion engines used in transportation systems such as automobiles, trucks, airplanes, boats and trains. Apart from the products derived from reactions 1 and 2, the gaseous effluents contain also the reagents fed in excess (oxygen and nitrogen, from atmospheric air), compounds originated from the impurities existing in fossil fuels (such as sulfur dioxide from the sulfur and airborne particulate from the ashes usually contained in fuels), and nitrogen oxides (nitrogen monoxide, nitrogen dioxide and nitrous oxide) resulting from chemical reactions taking place in high temperature flames between nitrogen and oxygen. Typically, the gaseous effluent from a coal fired power plant contains (expressed as volume in dry basis): 82% of nitrogen, 15% of carbon dioxide, 3% of oxygen, 1500 mg/m^3 of sulfur dioxide, 800 mg/m^3 of nitrogen oxides and 100 mg/m^3 of suspended particulate.

Since the ages the conquest of energy sources is what has made possible the growing development of civilization and industry as we know it today. For instance, the industrial revolution that took place in the end of the 18th century was only made possible by producing and controlling thermal energy in the form of steam, and the greater industrial development that took place after the beginning of the 20th century, was made possible by mastering production and transportation of electricity. In fact, the facility of access to the energy sources resulted in greater energy consumption in order to fulfill the demands of a growing world population particularly concerned with necessities in terms of comfort, well being and development which can be obtained by accessing modern and energy consuming technology and appliances. In fact, the International Energy Agency (IEA) predicts a 57% increase of the energy demand from 2004 to 2030 (IEA, 2007).

As the majority of energy is produced by the use of the combustion reactions previously mentioned, this energy consumption resulted in continuously growing emissions of carbon dioxide, which cannot be absorbed and compensated any more by natural mechanisms.

Table 1 shows comparisons on energy use, population and per capita consumption in 1900 and 2001 (Song, 2006). In the later years, over 85% of world energy demand is supplied by fossil fuels, and fossil-fueled power plants are responsible for about 40% of global CO_2 emissions, from which coal-fired power plants are the main contributor (Carapellucci and Milazzo, 2003). Apart from that, this table shows marked and significant differences between these two years: in these 100 years, the world population increases by 250%, the energy consumption increases around 915%, while the resulting carbon dioxide emissions increased by 1130%.

Table 1. World energy use, population and per capita in 1900 and 2001 (Song, 2006)

Index	1900	2001	Unit
Total energy consumption	991	10048	10^6 toe
Population	1762	6153	10^6
Energy consumption per capita	0.517	1633	Toe
Total CO_2 emissions	534×10^6	6607×10^6	ton C eqv
CO_2 emissions per capita	0.30	1.07	ton C eqv
CO_2 concentration in the atmosphere	295	371	ppm (v)
Life expectation	47.3	77.2	Years

Note: toe = ton of oil equivalent.

Of course, this resulted in a marked increase of CO_2 concentrations in the earth atmosphere from 295 ppm in 1900 to 315 ppm in 1960, when it started to be recorded by the Mauna-Loa observatory in Hawaii (previously, in the absence of accurate measurements of CO_2 concentrations in the atmosphere, the amounts are estimated by analysis of ice core data). Figures 1 and 2 show the evolution of CO_2 concentration in the earth atmosphere.

CO_2, unlikely to other atmospheric pollutants is not highly toxic, but is a greenhouse gas. CO_2 molecules that tend to accumulate in the earth atmosphere are going to create a "layer" which is effective in trapping the heat which is re-emitted from the earth surface due to the infra-red radiation received from the sun, causing difficulty in its dissipation, which is known as greenhouse effect. CO_2 is not the only gas responsible for creating this greenhouse effect as other gases molecules are able to contain infrared sun radiation more powerfully and, thus, are considered to have a higher heating potential, as shown in table 2. Figure 3 illustrates the mechanism of radiation entrapment by greenhouse gases (GHGs) molecules.

Although there is not an universal agreement on the cause, there is a growing consensus that global climate change is occurring and more and more climate scientists believe that a major cause is the anthropogenic emission of these greenhouse gases into the atmosphere (Figueroa *et al.*, 2008).

In spite of that, several scientists still believe that there is not a sound scientific basis to support this conclusion.

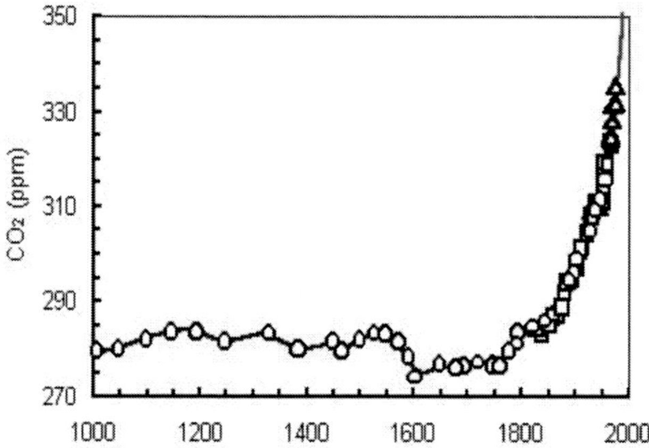

Source: CSIRO, 2007

Figure 1. Evolution of CO_2 concentration in earth atmosphere from 1000 to 2000.

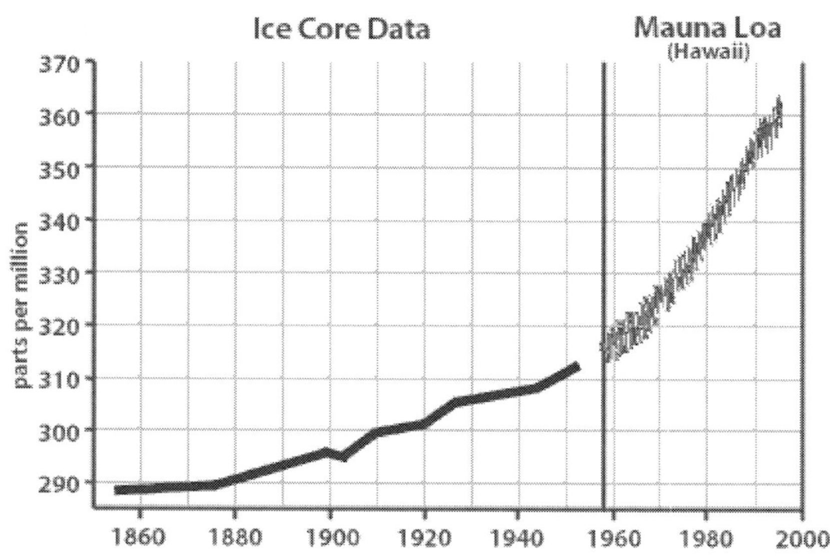

Source: Yang *et al.*, 2008.

Figure 2. Evolution of CO_2 concentration in earth atmosphere from 1860 to 2000.

Table 2. Heating potential and characteristics of greenhouse gases

Greenhouse gases	Pre-industrial concentration	Concentration in 1994	Atmos-pheric lifetime (years)	Anthropogenic sources	Global warming potential (CO_2 eqv)
CO_2 (carbon dioxide)	280 ppm (v)	358 ppm (v)	50-200	Fossil fuel combustion, land use conversion, cement production	1
CH_4 (methane)	700 ppb (v)	1720 ppm (v)	12-17	Fossil fuels, rice paddies, waste dumps, livestock	21
N_2O (nitrous oxide)	275 ppb (v)	312 ppm (v)	120-150	Fertilizer, industrial processes, combustion	310
CFC (chlorofluorocarbons)	0	503 ppt (v)	102	Liquid coolants, foams	125-152
HFC (hydrofluo-rocarbons)	0	105 ppt (v)	13	Liquid coolants	140-11700
PFC (perfluoro-carbons)	0	110 ppt (v)	50000	Production of aluminium	6500-9200
SF_6 (sulfur hexafluoride)	0	72 ppt (v)	1000	Production of magnesium	23900

Notes: ppt(v) = part per trillion by volume ; ppb(v) = part per billion by volume ; ppm(v) = part per million by volume.

Sources: IPCC radiative forcing report, 1995.

The science of climate change, UNEP and WMO, Cambridge University Press, 1996.

However, some irrefutable facts stand, such as the increase of the carbon dioxide concentration in the atmosphere, which is described in figures 1 and 2, and also that the global average temperature of the planet has been increasing since 1850, showing much more significant increases since 1970, as depicted in figure 4. This data suggests the existence of a correlation between the concentration of atmospheric carbon dioxide and the global average temperature, as shown in figure 5.

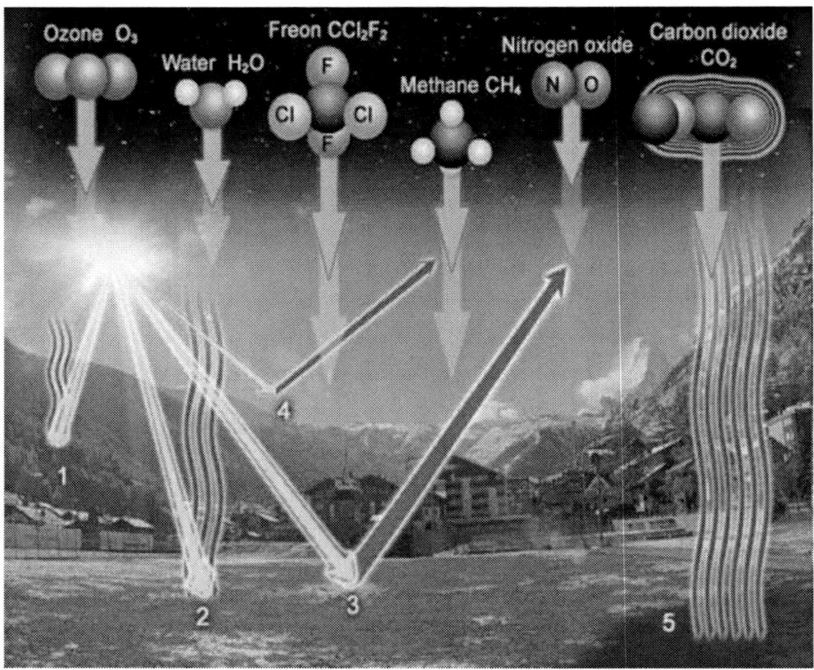

1: Solar radiation enters the earth atmosphere.
2: Solar radiation hits the earth surface.
3: Infrared radiation is reflected from earth surface.
4: Infrared radiation is trapped by GHGs molecules.
5: Infrared radiation is maintained within earth atmosphere.
Source: www.chemistryland.com.

Figure 3. Mechanism of radiation entrapment by GHGs molecules

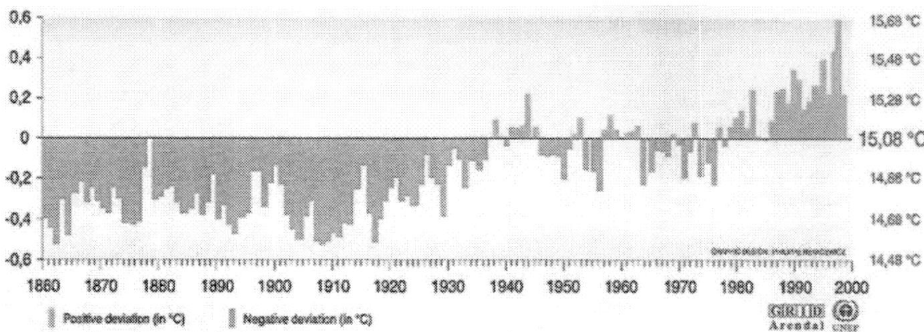

Source: http://maps.grida.no/go/graphic/trends-im-global-temperatures.

Figure 4. Evolution of global average temperature of the earth surface from 1860 until 2000.

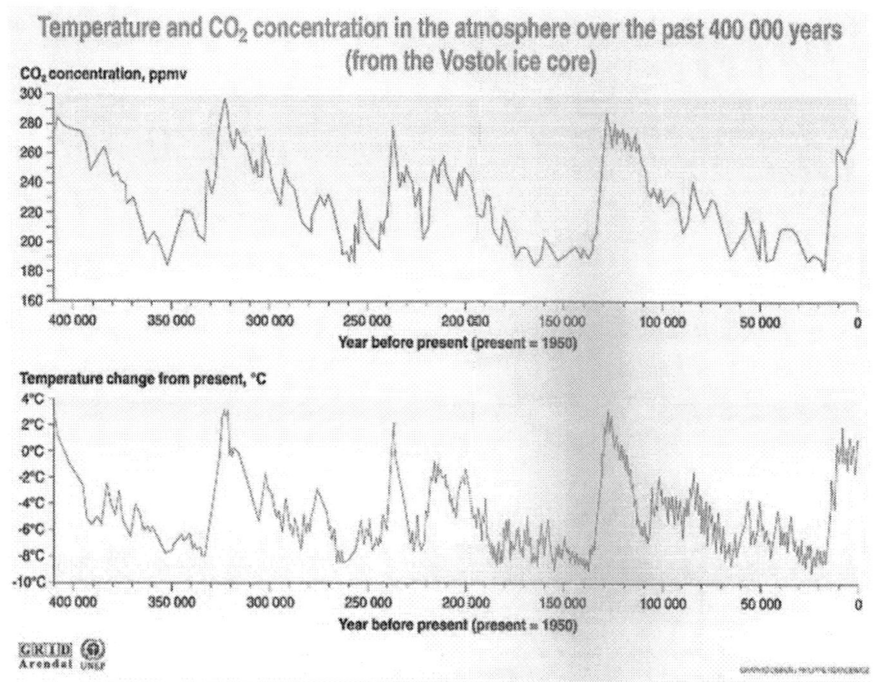

Source: http://maps.grida.no/go/graphic/temperature-and-co2-concentration-in-the-atmosphere-over-the-past-400-000 years.

Figure 5. Correlation of temperature variation and carbon dioxide concentration variation in the last 400,000 years.

Apparently the unanimity does not surpasses this point: the main debate lies within the cause of the observed heating, if this is due to the increase of anthropogenic carbon dioxide concentration or, on the other hand, it is merely the result of the natural variability of the earth's climate.

The Intergovernmental Panel on Climate Change (IPCC), which is a body created by the United Nations for studying these questions, claims that the climatic system of the earth has been altered significantly, globally and regionally, since the pre-industrial ages and the most important reason for these climatic changes seems to be the increase of GHGs in the atmosphere. The IPPC declared that the warming up of the climatic system of the earth cannot be denied, which has been expressed in several reports issued for more than 20 years. This opinion is based on evidences on the increases of the global average air temperature, global average sea temperature, defrosting of snow and ice and the average sea level. Thus, the major contribution for this heating phenomena occurring since the 20th century is estimated as 90% probability to anthropogenic activity.

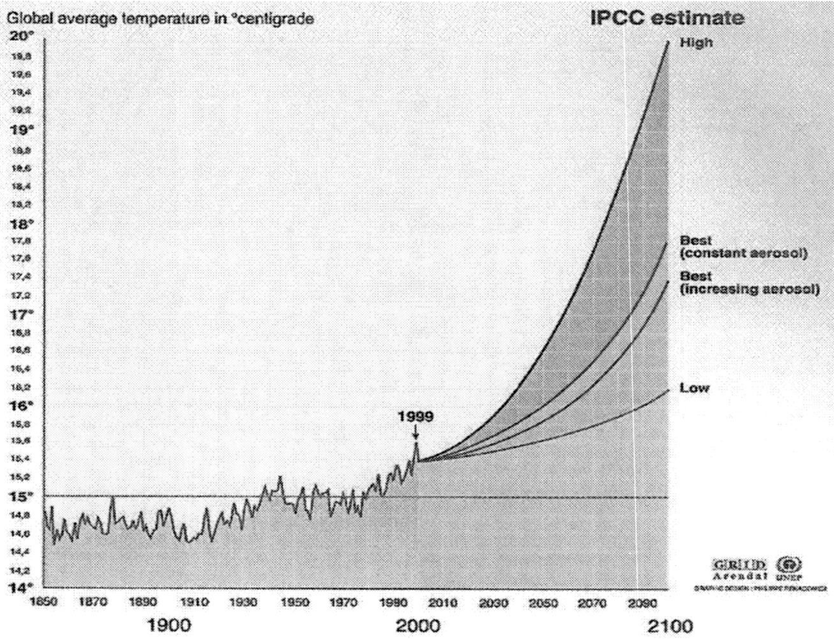

Source: http://maps.grida.no/go/graphic/temperature-trends-and-projections1.

Figure 6. Projected changes in global temperature.

Skeptics claim that the remaining 10% probability is not neglectable and that, in fact, the observed changes are mainly due to natural variations of the climatic system.

The climate theory is based on the assumption that the Earth maintains the equilibrium between the received solar radiation as ultraviolet and visible radiations which are absorbed and the infrared radiation which is reflected again to the space. Among all factors which are bound to affect this balance, the carbon dioxide concentration is the only one that is varying significantly. In fact, the increase on CO_2 concentration which was about 280 ppm (v) in 1750 and is now 380 ppm (v) is estimated, by IPPC, to surpass the natural range of 180-300 ppm (v) which was to be expected based on variations of the average historic concentrations dating from 800,000 years. Apart from that, the CO_2 increasing rate accounts for only 55% of the estimated emissions as a result of anthropogenic activities, which means that natural mechanisms are still able to sequestrate about 50% of the CO_2 generated by humans. Actually, it is estimated that oceans contain about 38,000 Gton of CO_2 (expressed as carbon), and are able to retain 1.7 Gton/year of CO_2 (expressed as carbon) from the atmosphere, while forests capture 1.4 Gton/year of CO_2 (expressed as carbon) (Yamasaki, 2003).

Of course, there is still a considerable debate going on about the consequences of the increase of the earth temperature. Figure 6 shows some projected changes in global temperature from global averages in the period 1856-1999 and previewed levels until 2100 according to different scenarios.

One possible and dramatic consequence is the sea level rise which is shown in figure 7 again resulting from taking the scenarios for changes in global temperature into consideration.

As a conclusion, and in spite of the existing controversy, the prevalent opinion seems to be that the set of existing data on climate points out to an undoubtedly evidence of the influence of human activities on climate change. Nevertheless, there are still questions and uncertainty about the projected and future consequences to be observed.

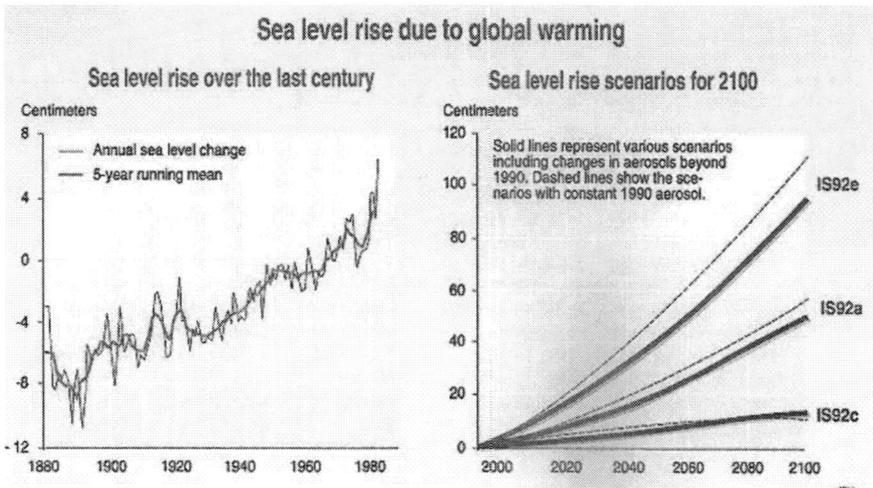

Sources: Climate change, UNEP and WMO, Cambridge University Press, 1995. Sea level rise over the last century, Gormitz and Lebedeff, 1987.

Figure 7. Projected sea level rise due to global warming.

Chapter 2

TECHNOLOGIES FOR CO_2 SEQUESTRATION

The previous chapter described some of the possible consequences of increasing CO_2 levels in the earth atmosphere. This chapter describes the available and emerging technologies for CO_2 capture and sequestration, which has been mainly based on a review by Figueroa et al. (2008), updated with recent developments.

In order to tackle this major problem there are mainly 3 different options, which were summarized by Y. Kaya and can be expressed as:

$$CO_2\uparrow = POP \times (GDP/POP) \times (BTU/GDP) \times (CO_2\uparrow/BTU) - CO_2\downarrow$$

where $CO_2\uparrow$ is the total CO_2 released to the atmosphere, POP is population, GDP/POP is per capita gross domestic product and is a measure of the standard of living, BTU/GDP is energy consumption per unit of GDP and is a measure of energy intensity; $CO_2\uparrow$/BTU is the amount of CO_2 released per unit of energy consumed and is a measure of carbon intensity, and $CO_2\downarrow$ is the amount of CO_2 stored or sequestered in biosphere and geosphere sinks (Olajire, 2010).

This equation indicates the factors that can be altered in order to observe a reduction of CO_2 emissions to the atmosphere. From those, reducing the population or the standard of living is not likely to be considered. Thus, the only three remaining options are as follows:

i) reducing energy intensity (BTU/GDP), which requires efficient use of energy;

ii) reducing carbon intensity ($CO_2\uparrow$/BTU), which implies the use of carbon-free fuel, and requires switching from traditional fuels to using non-fossil fuels such as hydrogen and renewable energy;
iii) enhancing the sequestration of CO_2, which involves the development of technologies to capture and sequester more and more CO_2.

It is clear that implementing all of the above mentioned options will be necessary if CO_2 emission abatement becomes a top-priority. However, at the current state of development, and the levels of risks and cost of non-fossil fuel energy alternatives such as nuclear, biomass, solar energy, etc., these energy sources cannot meet our energy demand currently fed by fossil fuels. Additionally, any rapid change to non-fossil energy sources, even if possible, would result in large disruptions to the existing energy supply infrastructure with substantial consequences to the global economy (Olajire, 2010).

Therefore, in order to meet mid to long-term CO_2 reduction targets, cost-effective CO_2 capture from fossil fuel uses and subsequent sequestration options need to be carefully evaluated, in view of the growing world demand for energy. To enhance the sequestration of CO_2, carbon dioxide capture and storage (CCS) technologies must be improved based on the development of new approaches of CO_2 separation and capture. Active CCS technologies require carbon emissions to be captured and stored in a form or location that is isolated from the atmosphere on a millennial time scale.

Capturing CO_2 from flue gas streams is an essential factor for carbon management for sequestration of CO_2 from our environment. Nowadays, the majority of energy generated in the world is obtained from the combustion of fossil fuels (mainly coal and natural gas) in power plants: it is estimated that about 85% of the world energy demand is supplied by fossil fuels. This is, in fact, due to the acceptable cost of fuels and the existence of reliable technology for energy production.

The US Department of Energy (DOE) previews that the consumption of fossil fuels (fueloil, coal and natural gas) will increase by 27% in the next 20 years, which will result in an increase of CO_2 emissions from 6,000 to 8,000 million tons per year in the USA by 2030 (Figueroa *et al.*, 2008). Even though, DOE estimates that, by 2030, the combined CO_2 emissions of China and India, from coal combustion, will be about three times more than the USA. China is expected to emit 8286 million tons, India 1371 million tons and USA 3226 million tons. This clearly demonstrates that no single nation can reduce the CO_2 emissions enough to stabilize GHGs concentrations in the atmosphere but, on the opposite site, the effort has to be unified and obtained at an acceptable cost in order to be compatible with a sustainable national and global economic growth (EIA, 2006).

Currently, about 60% of CO_2 is released from stationary sources, from thermal power plants. From these emissions, CO_2 content ranges from 8% in natural gas fueled power plants to 14% in coal fueled power plants (Panachi and Skogestad, 2010 ; Greer *et al.*, 2010).

Therefore, an important component of DOE's Carbon Sequestration Program is currently directed towards the reduction of CO_2 emissions from power plants. In fact, around one third of the antrophogenic CO_2 emissions in the USA is due to the operation of power plants. CO_2 emissions due to coal combustion, almost exclusively used for producing electricity, increased more than 18% in the period 1990-2003, and it is still expected to increase by 54% until 2030, if no specific measures are taken for CO_2 abatement. If the capture of CO_2 from power plants is, definitively a reduction option, then research and development on this subject is crucial in order to attain large scale implementation having economic and environmental acceptable impacts (Figueroa *et al.*, 2008).

Figure 8 shows a preview for the utilization of different fuels for producing electricity in the world: it is claimed that coal will continue to be the main fuel being used for electricity production in the near future. Therefore, there is an urgent need to adopt technologies which can perform the use of fossil fuels in a cleaner way, thus allowing for the start-up of a new greener and sustainable economy, in the future (Kothandaraman, 2010).

Like in the USA, world economies are still growing, which means that electricity demand will continue to increase too. EIA estimates that, in the USA, electricity demand will increase by 40% in the next 25 years.

The main advantage of adopting carbon capture and sequestration (CCS) techniques are that they may still allow that fossil fuels will continue to be used without being a major contribution towards global warming. This represents a rupture with the conventional approach on the mitigation of climatic changes, which calls for the elimination or limitation on the use of fossil fuels. As there is a considerable dependency on fossil fuels and there are also several difficulties (technical, economic and social) on the use of large scale alternative options (such as renewable and nuclear energies), the effort should be made on the use of fossil fuels that are capable of generate emissions that can be efficiently reduced. Of course that, coal fueled power plants which, as previously mentioned, contribute to 30% of CO_2 emissions in the USA, are the main target for the use of this type of technology (Rao and Rubin, 2009).

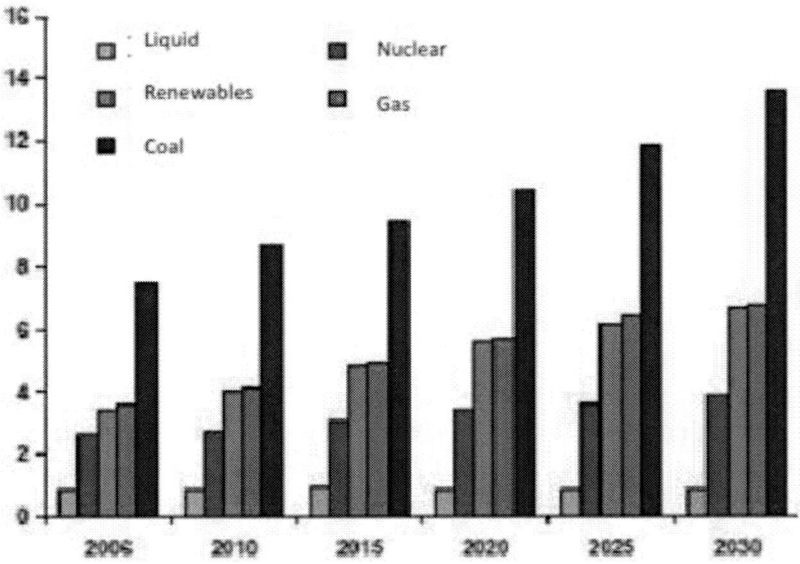

Source: (Kothandaraman, 2010).

Figure 8. Electricity production (Trillions kWh) in the world by fuel source, 2006-2030.

Once the CO_2 is captured, from its high emitting stationary sources, such as thermal power plants, it needs to be safely stored. Again, there are several options that can be taken, but the distance for a safe storing site, as well as its availability, and the transportation cost limits the chosen option. In general, technical studies indicate that deep geologic formations are the most attractive option for power plants. This approach will allow for a safe storage of CO_2 during several thousand years. However, even if the economic costs of this type of CO_2 storage are not excessive, the social and political acceptability are not yet clear.

CO_2 sequestration in geologic formations shows great promise for the large number of potential geologic sinks. Litynski *et al.* (2006) have estimated that 1120 to 3400 billion ton of CO_2 can be sequestered in the formations identified so far. Also, with higher petroleum prices, there is an increased interest in using CO_2 flooding as a means to enhance oil recovery (EOR), and with higher gas prices, there will be growing interest in using CO_2 for enhanced coal bed methane production (ECBM). However, none of these activities will be possible unless CO_2 is first captured. None of the current available CO_2 capture processes are economically feasible on a national implementation scale to capture CO_2 for sequestration, since they consume large amounts of parasitic power and significantly increase the cost of electricity. Thus, improved CO_2 capture technologies are vital if the promise of geologic sequestration, EOR and ECBM is to be realized (Figueroa *et al.*, 2008).

In consideration of how best to improve CO_2 capture, there are three technological approaches, shown in figure 9, which can be pursued for CO_2 capture from coal fuel power generation:

i. post-combustion capture;
ii. pre-combustion capture;
iii. oxy-combustion.

A novel process that will be described thereafter is the chemical-looping combustion process.

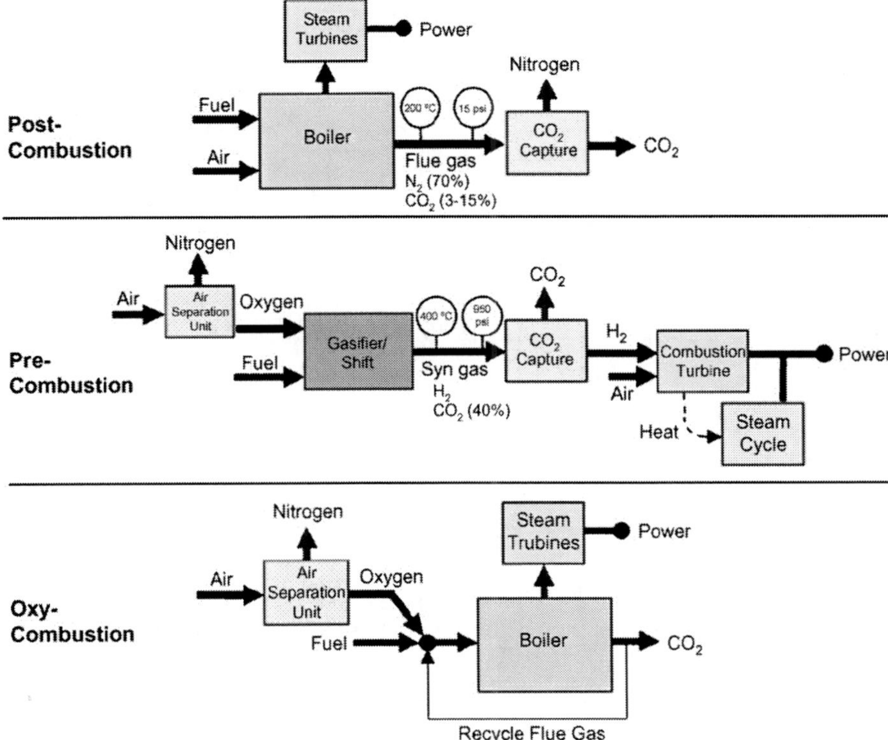

Figure 9. Main technical approaches for CO_2 capture (Figueroa *et al.*, 2008).

In the post-combustion process, CO_2 is separated from a flue gas environment containing NOx and SO_2. Figure 10 also shows this process flow (Feron and Hendriks, 2005). One way to do post-combustion capture is to use chemical absorption, such as monoethanolamine (MEA) absorption. This technique has been widely used in natural gas industry for over 60 years. Another advantage of this technique is that it produces a relatively pure CO_2 stream. Although the size and cost of the required absorber would be comparable to those of an SO_2 scrubber, the absorber would consume one-quart to one-third of the total steam produced by the plant, reducing its generating capacity by the same amount. Also the footprint of the host plant is increased by 60% (Elwell and Grant, 2006). The National Energy Technology Laboratory (NETL) estimated, in 2000, that this scheme would increase the cost of electricity production by 70% (Elwell and Grant, 2006). An alternative post-combustion CO_2 capture method is the use of membrane gas separation technologies.

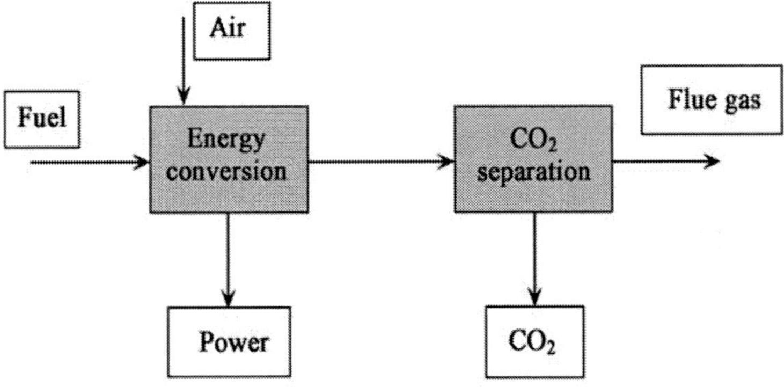

Figure 10. Pulverized coal combustion CO2 capture (Feron and Hendriks, 2005).

In the pre-combustion process, fuels are first converted into a mixture of CO_2 and H_2 through a reforming (natural gas) or gasification (coal) process and the subsequent shift-reaction. CO_2 can be separated from the conversion product stream and H_2 can then be burned in gas turbine or be used by fuel cell. Figure 11 shows this process flow (Feron and Hendriks, 2005). Gasification partially oxidizes coal to produce a gaseous fuel, which is essentially a hydrogen and carbon monoxide mixture. When syngas is used to fuel a plant similar to a traditional combined-cycle power plant, the process is referred to as integrated gasification combined-cycle (IGCC). Several methods can be used to capture CO_2. The leading option for CO_2 capture is an absorption process, in which the solvent can be a chemical one, such as MEA absorption process, or a physical one, such as pressure swing adsorption (PSA). Physical absorption is a mature technology and has been used in the Great Plains Synfuels plant in North Dakota, USA for 20 years (Elwell and Grant, 2006). This synfuel plant consume lignite coal and produces synthetic natural gas. Membrane technology has been used in the FutureGen power plant (Elwell and Grant, 2006). Pre-combustion capture is potentially less expensive than post-combustion capture. IGCC power plants applying pre-combustion capture are more efficient than pulverized coal-fired plants and would be the choice for new plants. The US Department of Energy (DOE) estimated in 2000 that the use of pre-combustion capture would increase the cost of electricity production by 25%. The long-term goal is to reduce the production penalty to 10% (Elwell and Grant, 2006).

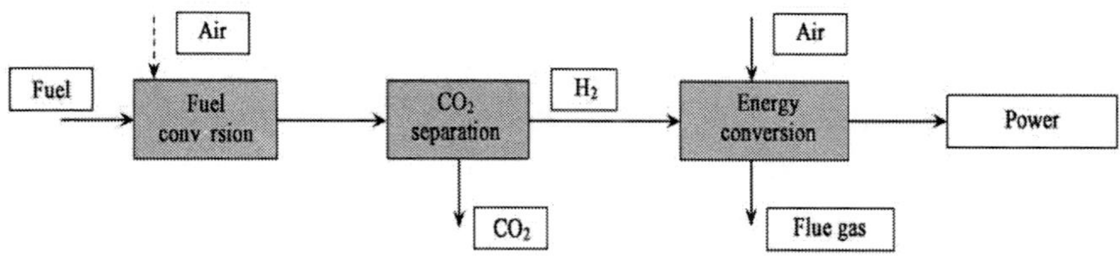

Figure 11. Gasification CO_2 capture process (Feron and Hendriks, 2005).

In the oxy-firing process, pure O_2 is separated from air and sent to energy conversion unit and combines with partially recycled flue gas of concentrated CO_2 to keep the furnace temperature below the allowable point. The combustion takes place in an environment of O_2/CO_2 mixture. The resulting flue gas is high-purity CO_2 stream. This process flow is shown in figure 12 (Feron and Hendriks, 2005). The exhaust gas stream is free of nitrogen components. Particulates and sulfur compounds are first removed from the exhaust stream using widely adopted techniques. After SO_2 removal, the exhaust gas stream is approximately 90% CO_2 by volume on a dry basis. Further separation of CO_2 is not necessary. CO_2 can then be compressed for storage or transportation. The main advantage of this technology is the elimination of NOx control equipment and the CO_2 separation step. The boiler size can be reduced because only oxygen is supplied for combustion. The size of subsequent equipment, such as SO_2 scrubber, can also be reduced. The disadvantage is the corrosion of equipment by increased SO_2 concentration in the exhaust gas stream. This technology is not mature and its operating and maintenance and capital costs would be comparable to that of post-combustion technology. The capital cost is estimated to be USD $2,040/kW (Elwell and Grant, 2006). Specifically, the oxygen separation plant would consume about 23% to 37% of the total plant output and costs about the same as a chemical absorber.

Table 3 provides a summary of the inherent advantages and disadvantages of each of these pathways. Post-combustion capture applies primarily to coal-fueled power generators that are air fired. Pre-combustion capture applies to gasification plants. Oxy-combustion can be applied to new plants or retrofitted to existing plants. Figure 13 indicates that as innovative CO_2 capture and separation technologies advance significant cost reduction benefits can potentially be realized once they are commercialized, and Figure 14 presents the main technology options for CO_2 capture that will be described thereafter.

Chemical-looping combustion (CLC) is a novel process with inherent CO_2 capture. It has also been called unmixed combustion since direct contact between fuel and combustion air is avoided. Instead, an oxygen carrier brings oxygen from air to fuel. Suitable oxygen carriers are small particles of metal oxide such as Fe_2O_3, NiO, CuO or Mn_2O_3. A basic CLC system is shown in Figure 15 (Rydén and Lyngfelt, 2006). The CLC has two reactors, one for air and one for fuel. The oxygen carrier circulates between the reactors. In the air reactor, the carrier is oxidized by oxygen according the following first reaction. In the fuel reactor, the metal oxide is reduced by fuel, which is oxidized to CO_2 and H_2O, according to the following reactions:

$$O_2 + 2Me \rightarrow 2MeO$$
$$C_nH_m + (2n + \tfrac{1}{2}m)MeO \rightarrow nCO_2 + \tfrac{1}{2}mH_2O + (2n + \tfrac{1}{2}m)Me$$

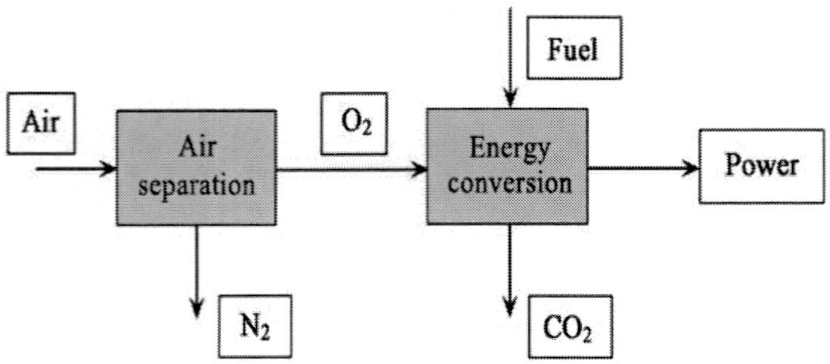

Figure 12. Oxy-firing CO_2 capture process (Feron and Hendriks, 2005).

Table 3. Advantages of the main technical approaches for CO2 capture (Figueroa et al., 2008)

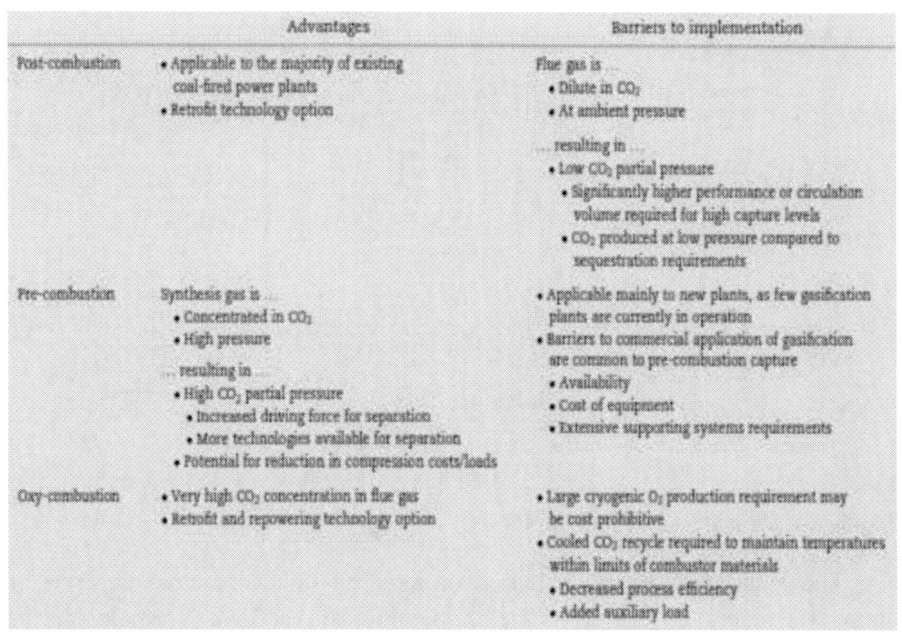

	Advantages	Barriers to implementation
Post-combustion	• Applicable to the majority of existing coal-fired power plants • Retrofit technology option	Flue gas is … • Dilute in CO_2 • At ambient pressure … resulting in … • Low CO_2 partial pressure • Significantly higher performance or circulation volume required for high capture levels • CO_2 produced at low pressure compared to sequestration requirements
Pre-combustion	Synthesis gas is … • Concentrated in CO_2 • High pressure … resulting in … • High CO_2 partial pressure • Increased driving force for separation • More technologies available for separation • Potential for reduction in compression costs/loads	• Applicable mainly to new plants, as few gasification plants are currently in operation • Barriers to commercial application of gasification are common to pre-combustion capture • Availability • Cost of equipment • Extensive supporting systems requirements
Oxy-combustion	• Very high CO_2 concentration in flue gas • Retrofit and repowering technology option	• Large cryogenic O_2 production requirement may be cost prohibitive • Cooled CO_2 recycle required to maintain temperatures within limits of combustor materials • Decreased process efficiency • Added auxiliary load

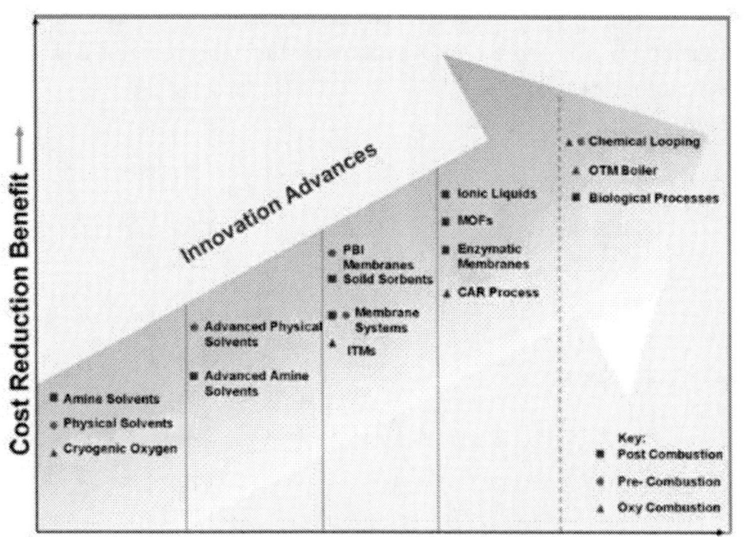

Figure 13. Innovative technologies for CO_2 capture: benefits in terms of reduction costs versus commercialization time (Figueroa *et al.*, 2008).

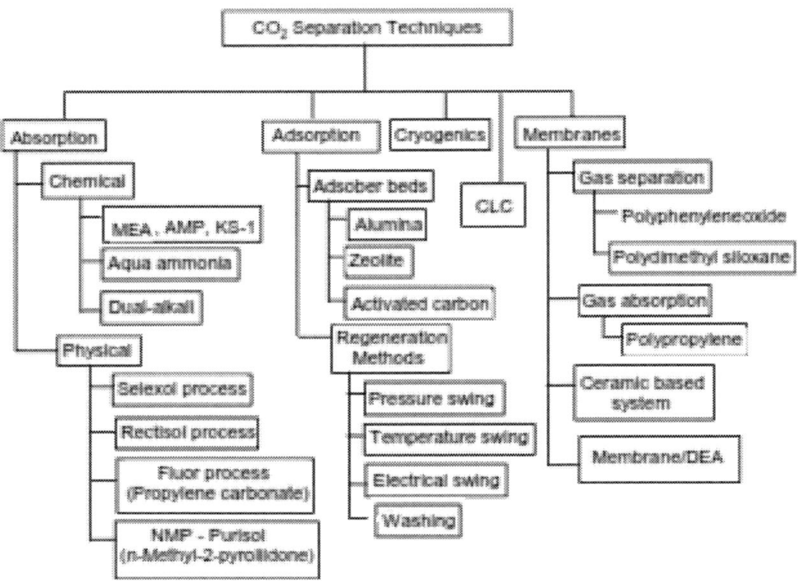

Figure 14. Main technology option for CO_2 capture (Olajire, 2010).

The amount of energy released or required in the reactors depends on these two reactions, as well as the temperature of reactions. Normally, the first reaction is strongly exothermic, whereas the second reaction is usually endothermic but for some combinations of fuel and oxygen carrier it may be slightly exothermic.

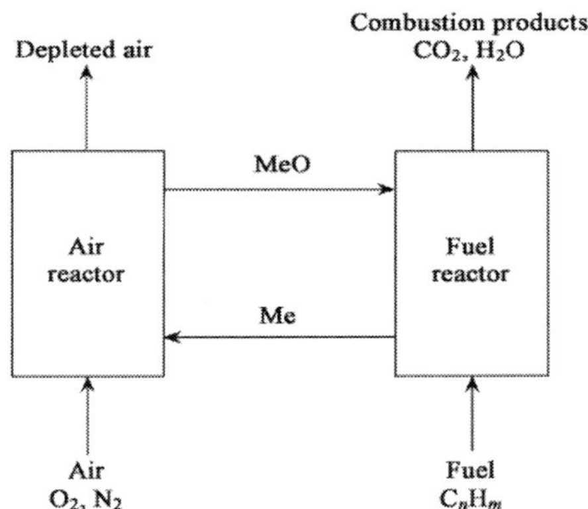

Figure 15. Chemical-looping combustion (Rydén and Lyngfelt, 2006).
Me: metal; MeO: metal oxide as oxygen carrier.

In principle, all major fuels can be utilized in a CLC. CLC has several advantages compared with convention- al combustion. The exhaust gas stream from the air reactor is harmless, consisting mainly of N_2. In a well-designed system, there should be no thermal formation of NOx since the regeneration of oxygen carrier takes place without flame and at moderate temperatures. The exhaust gas from the fuel reactor consists of CO_2 and H_2O. Separation of CO_2 can be done by a condenser. This is the major advantage with CLC which avoids the huge energy penalty necessary in traditional amine scrubbing process to capture CO_2.

From what was mentioned previously, it becomes clear that, for previously existing combustion facilities, post- combustion capture is the most feasible alternative, as it involves the removal of CO_2 from the flue gas produced by combustion. Existing power plants use air, which is almost 80% of nitrogen, for combustion and generate a flue gas that is nearly at atmospheric pressure, and, typically, has a CO_2 concentration of less than 15%. Thus, the thermodynamic driving force for CO_2 capture from flue gas is low (CO_2 partial pressure is typically less than 0.15 atm) creating a technical challenge for the development of cost effective advanced capture processes. In spite of this difficulty, post-combustion carbon capture has the greatest near-term potential for reducing greenhouse gas emissions, because it can be retrofitted to existing units that generate around 30% of the CO_2 emissions in the power generation sector.

2.1. Carbon Fixation by Natural Processes

There are both natural and artificial ways to capture or fix the carbon to avoid (or delay) emission into the atmosphere, such as forestation, ocean fertilization, photosynthesis, mineral carbonation, and hydrate (Figueroa, 2008). Geological injection and direct ocean dump are not covered.

2.1.1. Forestation

It is estimated that 1.4±0.7 Gt-carbon is captured by terrestrial systems from atmosphere via photosynthesis (Yamasaki, 2003). Increasing this carbon flux is achieved by forestation, reforestation of arid lands and greening of deserts. At the beginning of this sequestration, the carbon flow to terrestrial vegetation is positive and the "stock" of carbon is increased. When fully developed, the net capture becomes zero because of the balance of capturing and release. The potential capacity for carbon sequestration in terrestrial systems is estimated to be 5-10 Gt-carbon annually (Yamasaki, 2003).

2.1.2. Ocean Fertilization

Ocean stores more CO_2 than terrestrial vegetation. The ocean contains about 38,000 Gt-carbon, and about 1.7±0.5 Gt is taken up annually from the atmosphere (Yamasaki, 2003). The production of phytoplankton at 50–100 Gt carbon annually is much higher than that of terrestrial vegetation (Yamasaki, 2003). Part of the carbon would be released back into the atmosphere by respiration process, and the remaining would descend into the deeper ocean in the form of particulate organic matter either by the death of phytoplankton or after grazing. This sequestration process could be enhanced by ocean fertilization. Ocean fertilization refers to the practice of increasing limiting nutrients to stimulate the production of phytoplankton. Ocean fertilization has been challenged to interfere with the marine ecosystem which could lead to fatal impact. In addition, the decomposition of sinking organic matter could produce other stronger greenhouse gases, such as methane or nitrogen monoxide. This method needs to be further studied before it is put into practice.

2.1.3. Photosynthesis Process

Most of the current CCS options offer short- to mid-term solutions with associated drawbacks. For instance, geologic and oceanic injection only delays the release of CO_2 into atmosphere. Authors believe that only by switching to renewable fuels can solve the CO_2 emission problem. Biological carbon sequestration using technologies such as controlled photosynthetic reactions may help to alleviate GHG problems in a sustainable way. Stewart *et al.* studied a system that combines solar energy collection and delivery through a fiber optic system to stimulate the growth of biological organisms in a bio-generator to produce useful by-products from carbon dioxide (Stewart and Hessami, 2005). The photo-bioreactor system makes use of the natural photosynthesis process to convert light, heat and carbon dioxide to useful products, such as carbohydrates, hydrogen and oxygen. The type of product depends on the biological strains used in the photo-bioreactor. The following equation illustrates a photosynthetic reaction:

$$6CO_2(aq) + 6H_2O(l) + light + heat \rightarrow C_6H_{12}O_6(aq) + 6O_2(g)$$

Cyanobacteria or micro-algae have been suggested to perform the role of photosynthesis. In order to promote uniform growth of the organisms, the distribution of light flux in the wavelength range of 400–700 nm needs to be delivered to the bioreactor. The efficient distribution of light throughout the photo-bioreactor will affect the carbon dioxide uptake rates. Despite 50 years of development of closed-system photo-bioreactor systems, commercial viability has not yet been achieved. At present, open pond systems produce around 100 t of biomass annually at a cost of around USD $10,000/metric ton (Stewart and Hessami, 2005).

2.1.4. Mineral Carbonation, Natural or Biomimetic

Another major CO_2 fixation process in nature is the chemical weathering of rocks, such as silicates, containing calcium or magnesium. The silicate rocks could be turned into carbonates by reacting with CO_2 following this mechanism (Maroto-Valera *et al.*, 2005):

$$(Mg,Ca)_x Si_y O_{x+2y} + xCO_2 \rightarrow x(Mg,Ca)CO_3 + ySiO_2$$

Mineral carbonation results in the storage of CO_2 in solid form as a stable, environmentally benign mineral carbonate. The energy state of mineral carbonate is 60 to 180 kJ/mol lower than CO_2, which is 400 kJ/mol lower than carbon (Maroto-Valera *et al.*, 2005). Consequently, sequestration in the form of a carbonate ensures long-term fixation of CO_2 rather that temporary storage. The risk of any accidental release of stored CO_2 leaking from underground can also be avoided.

The weathering of rocks happens in nature over geologic time scales. Artificial ways were first proposed by Seifritz in 1990 to emulate this natural process (Liu *et al.*, 2005). Two methods are possible to accelerate the reaction rates, i.e., to enhance the dissolution of mineral ions using stronger acids, or to react under higher CO_2 pressure. This artificial process is called CO_2 mineralization, or mineral carbonation. Many researchers studied the mineral carbonation process (Druckenmiller and Maroto-Valera, 2005; Haywood *et al.*, 2001; Liu *et al.*, 2005; Maroto-Valera *et al.*, 2005; Stolaroff *et al.*, 2005). The capacity of this fixation is huge, even though the use of acid may cause problems in both the reaction facilities and in the environment, and higher CO_2 pressure consumes intensive energy. The carbonation of silicate minerals by mechanochemical process was investigated by University of Utah and University of Idaho (Nelson, 2004; Nelson and Prisbrey, 2004). Natural and synthetic silicate compounds were ground for a range of times in the presence of gaseous CO_2. Studied materials include naturally-occurring materials, such as forsterite, lizardite, and wollastonite. Synthetic materials such as magnesium silicate were also studied. It was shown that a significant change takes place in the lizardite variety of serpentine after 15 to 20 min of intense grinding in the presence of gaseous CO_2. Analyses showed that small amounts of carbon are fixed by grinding lizardite, forsterite, and wollastonite, and magnesium silicate, in the presence of gaseous CO2. Considering the energy input to grind the materials to a size that can fix CO_2, this technology has limited use in practice. Maroto-Valera *et al.* studied mineral carbonation through surface activation to promote and accelerate carbonation reaction rates and effciencies at moderate reaction conditions (Maroto-Valera *et al.*, 2005). Their results indicate that the surface area, of the raw serpentine, can be significantly increased through physical and chemical activation methods. The chemical activations were more effective than physical activation at increasing the surface area. Sulfuric acid was found to be the most effective acid used during the chemical activation. The steam activation promotes the CO_2 capture by serpentine. The most promising results came from the carbonation of the extracted $Mg(OH)_2$ solution, where the carbonation efficiency was estimated to be at least 53 percent. Liu *et al.* developed a biomimetic process, in which a biological catalyst, the enzyme carbonic anhydrase, was used to accelerate an aqueous processing route to carbonate formation (Liu *et al.*, 2005). The enzyme was immobilized on chitosan-alginate beads. Simulated brine containing calcium was used to provide cations for carbonate formation. Results show that carbonic anhydrase is a promising enzyme to increase the rate of precipitation of carbonate mineral, with industrial wastewater as a cation source. According to a study performed by Haywood *et al.*, without any radical alteration of the various process routes, it seems unlikely that CO_2 sequestration within a stable mineral carbonate is a realistic proposal (Haywood *et al.*, 2001). Widespread application of this technology will require a huge supply of metal oxides enough to cause major impacts upon iron, manganese, chromium and nickel mining industries.

2.1.5. In-situ CO_2 Capture

Hughes *et al.* explored the *in-situ* CO_2 capture using carbonation mechanism at high temperature (700 °C) at atmospheric pressure in a dual fluidized bed combustion system (Hughes *et al.*, 2005). The following reactions take place inside the reactors:

$$CaO + CO_2 \rightarrow CaCO_3$$

Similar to CLC, there are two reactors in this system. Primary fuel combustion takes place in the carbonator and sorbent CaO reacts with CO_2 in the same reactor achieving *in-situ* CO_2 capture. $CaCO_3$ is heated to regenerate CaO in the calciner and the heat is provided by burning a secondary fuel such as petroleum coke. The overall thermal efficiency was shown to be comparable to that of a current combustion system without CO_2 capture. CANMET has extensively studied similar approaches (Sun *et al.*, 2007; Manovic and Anthony, 2007; MacKenzie *et al.*, 2007).

2.1.6. Hydrate-based Separations

Hydrate-based separation is also a promising. This process starts by forming the hydrate, and thus, by exposing the exhaust gas streaming containing CO_2 to water under high pressure. As long as the hydrate is formed, the CO_2 is captured. The hydrate is then separated and dissociated, thereby releasing CO_2. This technology has an energy penalty as small as 7% (Elwell and Grant, 2006).

2.2. Chemical Absorption by Amine Based Systems

Amines react with CO_2 to form water soluble compounds. Because of this, amines are able to capture CO_2 from streams having a low CO_2 partial pressure, but its capacity is equilibrium limited. Therefore, amine-based systems are able to recover CO_2 from the flue gases of conventional coal fired power plants, however with a significant cost and efficiency penalty. Although amines have been used for many years, particularly in the removal of sour gases from natural fuel gas, there is still room for process improvement.

Amines are, actually, available in three forms (primary, secondary and tertiary), each having its advantages and disadvantages as a CO_2 solvent. In addition to the amines, additives can be used to increase system performance. Finally, design modifications are possible in order to decrease capital costs and improve energy and process integration.

Monoethanolamine (MEA) is the most used amine, as a solvent for CO_2 capture, and this is the system for which more thermodynamic data is available. The compound's formula is $CH_2(NH_2)CH_2OH$, which means that it behaves like a primary alcohol (due to the hydroxyl group) and also as a primary amine (due to the amine group), as shown in figure 16.

Figure 16. Molecular structure of MEA.

MEA is widely used for CO_2 removal from natural gas streams as well as refinery processes streams. MEA is a strong alkali, reacting quickly and producing a low CO_2 concentration. The solubility and reaction kinetics of CO_2 in aqueous solutions of MEA have also been studied: even if MEA reacts quickly with CO_2, the absorption ratio is controlled by the reaction kinetics, and the absorption efficiency is usually lower than 20 %.

MEA is a relatively cheap solvent and the chemical absorption of CO_2 is actually supported by commercially available and tested technology, which means that alternative solvents must have higher CO_2 capture capacities and also lower energy consumptions.

The main disadvantages of using MEA, as a solvent for post-combustion CO_2 capture are the following:

- low absorption capacity per carbon unit;
- imply high energy consumptions;
- solvent degrades due to the presence of SO_2 and O_2 in combustion gases;
- creates high corrosion in equipment.

Other interesting amines for CO_2 absorption are piperazine (PZ), 1,6-hexanodiamine (HAD) and ethylenediamine (EDA). Piperazine (PZ), which is a heterocyclic saturated compound, has a molecular structure as shown in figure 17.

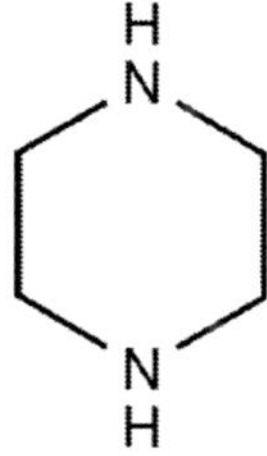

Figure 17. Molecular structure of piperazine (PZ).

Aqueous concentrated PZ has been studied as a new amine solvent for CO_2 absorption. The absorption rate of CO_2 in aqueous PZ is much faster than MEA. Another advantage is that aqueous PZ does not thermically degrades until 150 °C, and PZ has an energy requirement 10-20% lower than MEA. However, PZ has also some disadvantages as an absorption solvent as pure PZ is only soluble in water until 1.9 mol/kg of water.

1,6-hexanodiamine (HAD) is an amine with a hexametilene group connected to the amine functional groups in each end, having the chemical formula $NH_2(CH_2)_6NH_2$. Ethylenediamine (EDA) has the chemical formula $C_2H_4(NH_2)_2$ and its molecular structure is shown in figure 19.

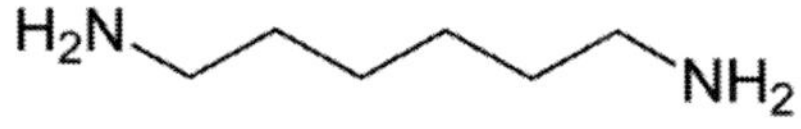

H_2N NH_2

Figure 18. Molecular structure of 1,6-hexanodiamine.

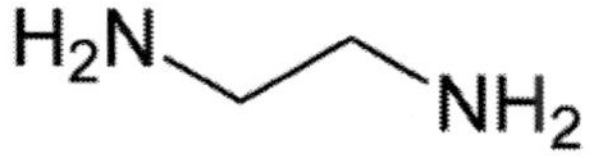

H_2N NH_2

Figure 19. Molecular structure of ethylenediamine.

Improvements to amine-based systems for post-combustion CO_2 capture are still being pursued by a number of researchers and process developers. Some of these are Fluor, Mitsubishi Heavy Industries (MHI) and Cansolv Technologies. Fluor´s Econamine FG Plus is a gas removal system that has demonstrated greater than 95% availability with natural gas fired power plants, such as a 350 ton/day CO_2 capture plant in Bellingham, MA, USA. This is, currently, a state-of-the-art commercial technology baseline and is used in comparison with other CO_2 capture technologies- MHI has developed a new absorption solvent, referred to as KS-1, A key factor in this development is the utilization of a new amine-type solvent for the capture of CO_2 from flue gas (BP America, 2005). Cansolv Technologies, Inc. proposes to reduce costs by incorporating CO_2 capture in a single column which also captures other pollutants, such as SO_2, NO_x and heavy metals. The new DC103 tertiary amine solvent has demonstrated fast mass transfer and good chemical stability with high capacity, around 0.5 mole CO_2/mole of amine per cycle compared to 0.25 mole/mole for monoethanolamine (MEA) (Hakka, 2007).

Apart from new amine solvent development, R&D pathways to improved amine-based systems include modified tower packing to reduce pressure drop and increase contacting, increased heat integration to reduce energy requirements, additives to reduce corrosion and allow higher amine concentrations, as well as improved regeneration procedures.

A typical amine solution process for CO2 capture is depicted in figure 20.

The system comprises three sections which are: i) absorption of CO_2 by an amine solution in an absorption column; ii) deabsorption of the amine rich solution in a stripping column; iii) heat exchangers.

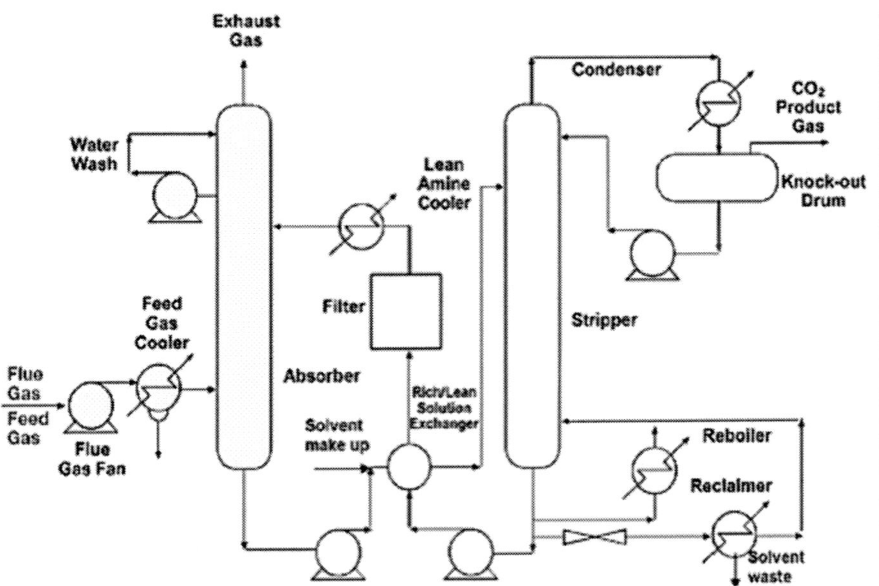

Figure 20. Flow sheet of a typical CO_2 capture process using amine solutions.

Prior to arriving at the first section, the combustion gases are cooled and compressed. In natural gas fueled power plants, the temperature of the flue gases is in the range 110-120 °C, which means that the gases need to be cooled before entering the absorption column, and the same happens with flue gases from coal fired power plants. Flue gases are, usually, cooled in a cooling tower, counter-current with water. Flue gases enter the lower part of the tower and the cooling water enters the top part of the tower, where the flue gases are cooled by water evaporation. Cooling water is collected on the base of the tower, and the cooled flue gases leave the tower by the top and, then, are sent to a blower where they are slightly compressed. Also, prior to entering the absorption column impurities such as SOx and NOx have to be removed.

In the first column, which is an absorbing column, CO_2 from flue gases is absorbed, at atmospheric pressure in a water solution of 30% (weight) of amine. The temperature inside the column is, usually, in the range 40-60 ºC, and the absorbing solution starts to heat up while CO_2 is being absorbed. The stream leaving the column is the CO_2 rich amine solution, with around 5.3% (mol) CO_2, which is fed to the second column, the stripping column. Stripping takes place at higher temperatures (100-120 ºC) and at pressure of 1.5-2 atm. Before entering the stripping column, this stream is pre-heated in a heat exchanger by the amine recycled solution (amine lean solution) with 2.3% (mol) CO_2. The pressure inside the stripping column is 2 bar. The amine rich solution is fed on the top of the stripping column and flows downward counter-current with the gas phase coming from a reboiler. The stream leaving the top of the stripping column is fed to a condenser and to a flash separator in order to separate CO_2 from water. Part of the reflux, formed by pure water, is send back to the top of the column; while a purge stream is send for storage. The energy consumed in the reboiler is needed to increase the temperature of the amine rich solution inside the stripping column, to promote the de-absorption reaction thus releasing CO_2, and also to produce steam in order to transfer CO_2 from the liquid to the gaseous phase. However, this is the most important heat loss in the whole process, which accounts for about 15-30% of the net power generated by a cola fired power plant.

The regenerated solvent solution (lean amine) is then pumped again to the absorption column, which passes through a heat exchanger to increase its temperature. Part of the product from the stripping column is send to a reclaimer unit where the solvent is vaporized and any impurities from volatile solvents. Gaseous CO_2 released from the stripping column needs to be dried up and be compressed before being send for safe storage.

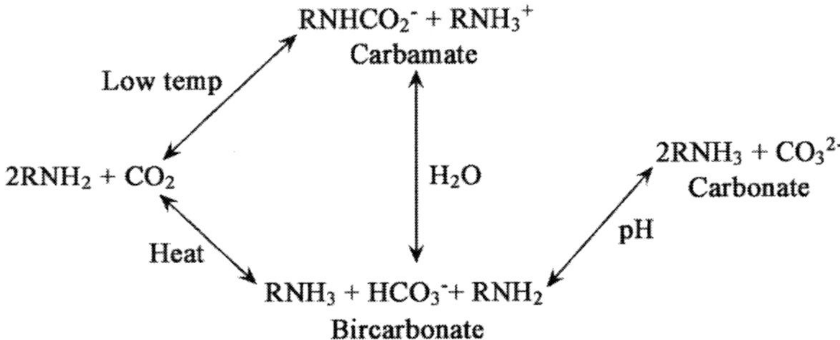

Figure 21. Proposed reaction sequence for the capture of carbon dioxide by liquid amine-based systems (Gray *et al.*, 2005).

Aminated compounds react with CO_2 to form stable carbamates or bicarbonate species, and this reaction can be reverted at high temperatures. The reaction scheme is shown in figure 21.

Figure 22 shows a detail of the stripping column which is the main energy consumer of this CO_2 capture system. The amine rich stream (S1) coming from the absorption column is pumped to the top of the stripping column, where it flows downward to the column filling, typically with a large superficial area (200 m^2/m^3).

The steam flow leaving the reboiler (S9) enters the stripping column in the base and goes upward in counter flow. From the rich liquid flow in base of the column, chemical reactions are reversed and release CO_2, which was ionically linked in aqueous solution, and is then diffused in the gaseous phase. The lean MEA solution leaves the lower part of the stripping column, and is then send to the absorption column (S8). Part of the lean solution is directed to the reboiler in order to generate steam to the regeneration process (S5). A condenser cools the outlet gas and, thus, reduces the water content, concentrating CO_2. The reboiler consumes the energy necessary to vaporize the liquid and also supplies the necessary energy to revert the chemical reactions. Usually the reboiler is heated by steam at approximately 120 ºC, which is the limit temperature at which the monoethanolamine starts to degrade. On the top of the stripping column, there is a condenser to cool the gas phase reaching the top of the column. Part of this steam condenses and is send back to the stripping column.

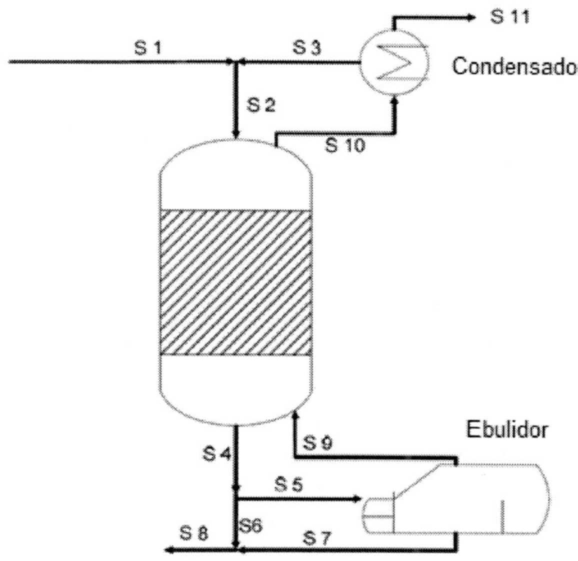

Figure 22. Flow diagram on the stripping column.

2.3. Chemical Absorption Using other Systems

Regarding absorption, other compounds, apart from amines, are used or are being currently tested as follows:

2.3.1. Carbonate-based Systems

Carbonate systems are based on the ability of a soluble carbonate to react with CO_2 to form bicarbonate, which when heated releases CO_2 and reverts to a carbonate. A major advantage of carbonates over amine-based systems is the significantly lower energy required for regeneration. The University of Texas at Austin has been developing a K_2CO_3 based system in which the solvent is promoted with catalytic amounts of piperazine (PZ). The K_2CO_3/PZ system has an absorption rate 10–30% faster than a 30% solution of MEA and also has favorable equilibrium characteristics. A benefit is that oxygen is less soluble in K^+/PZ solvents; however, piperazine is more expensive than MEA, so the economic impact of oxidative degradation will be about the same (Rochelle, 2006). Further analysis has indicated that the energy requirement is approximately 5% lower with a higher loading capacity of 40% versus about 30% for MEA, and system integration studies indicate that improvements in structured packing can provide an additional 5% energy savings, and multi-pressure stripping can reduce energy use 5–15% (Rochelle et al., 2006).

2.3.2. Aqueous Ammonia

Ammonia-based wet scrubbing is similar, in operation, to amine systems. Ammonia and its derivatives react with CO_2 via various mechanisms, one of which is the reaction of ammonium carbonate (AC) with CO_2, and water to form ammonium bicarbonate (ABC). This reaction has a significantly lower heat of reaction than amine-based systems, resulting in energy savings, provided the absorption/desorption cycle can be limited to this mechanism only. Ammonia-based absorption has a number of other advantages over amine-based systems, such as the potential for high CO_2 capacity, lack of degradation during absorption/regeneration, tolerance to oxygen in the flue gas, low cost, and potential for regeneration at high pressure. There is also the possibility of reaction with SO_x and NO_x to form fertilizers (ammonium sulfate and ammonium nitrate) as a sealable by-product. However, few concerns exist related to ammonia's higher volatility compared to that of MEA: the flue gas must be cooled to a range of 60–80 °C, to enhance the CO_2 absorptivity of the ammonia compounds and also to minimize ammonia vapor emissions during the absorption step. Additionally, there is some concern over ammonia losses during regeneration, which occurs at elevated temperatures. R&D process improvements include process optimization to increase CO_2 loading and use of various engineering techniques to eliminate ammonia vapor losses from the system during operation (Resnik *et al.*, 2004, 2006; Yeh *et al.*, 2005). Another ammonia-based system, under development by Alstom, is the chilled ammonia process (CAP), which performed a 5-MW pilot test in 2007 at We Energies Pleasant Prairie Power Plant. This process uses the same AC/ABC absorption chemistry as the aqueous system described above, but differs in that no fertilizer is produced and a slurry of aqueous AC/ABC, and solid ABC is circulated to capture CO_2 (Black, 2006). The process operates at near freezing temperatures (32–50 °C), and the flue gas is cooled prior to absorption using chilled water and a series of direct contact coolers. Technical barriers associated with the technology include cooling the flue gas and absorber to maintain operating temperatures below 50 ºC (required to reduce ammonia slip, achieve high CO_2 capacities, and for AC/ABC cycling), mitigating the ammonia slip during absorption and regeneration, achieving 90% removal efficiencies in a single stage, and avoiding fouling of heat transfer equipment by ABC deposition as a result of absorber operation with a saturated solution. Both the aqueous and chilled ammonia processes have the potential for improved energy efficiency over amine-based systems, if these technical barriers can be overcome.

2.3.3. Recent Research Programmes

This section seeks to give an overview of relevant important research programmes worldwide in post-combustion CO_2 capture, mainly based on a review by Wang et al. (2010).

2.3.3.1. Luminant Carbon Management Programme

The Luminant carbon management programme led by Prof. Gary Rochelle in the Department of Chemical Engineering, University of Texas at Austin (USA) focuses on tackling technical obstacles to the deployment of post-combustion CO_2 separation from flue gas by alkanolamine absorption/stripping and integrating the design of the capture process with aquifer storage/enhanced oil recovery processes (Rochelle, 2010).

The schematic of the pilot plant facility is shown in Figure 23. Both the absorber and stripper columns are packed columns with internal diameters of 0.427 m and total column height of 11 m. Columns consist of two 3.05 m packed bed sections with a collector plate and redistributor between the beds. Both random and structured packings were used alternatively in the two columns. The facility has a capacity to process approximately 3 ton of CO_2 per day (Dugas, 2006). The flue gas stream is prepared since it is not obtained directly from a power plant.

At this university pilot plant, several studies are carried out where research looks into CO_2 rate kinetics and solubility measurements (Bishnoi and Rochelle, 2002; Seibert *et al.*, 2011), degradation of solvents (Chi and Rochelle, 2002; Davis and Rochelle, 2009), systems modeling (Freguia and Rochelle, 2003; Ziaii *et al.*, 2009), pilot plant testing (Dugas, 2006; Chen *et al.*, 2006) among others. One of the main projects carried out was the carbon dioxide capture by absorption with potassium carbonate. This project ran from 2002 to 2007 with the aim to improve the process for CO_2 capture by alkanolamine absorption/stripping by developing an alternative solvent, aqueous potassium carbonate (K_2CO_3) promoted by piperazine (NETL,

2008). Accomplishments include evaluation of three solvents: MEA, and two variants of the piperazine-promoted K_2CO_3. It was shown that the requirements for one of the piperazine promoted K_2CO_3 solvents was much less than the conventional MEA due to increased absorption capacity and rates, as well as reduced heat of absorption, implying reduced regeneration requirements.

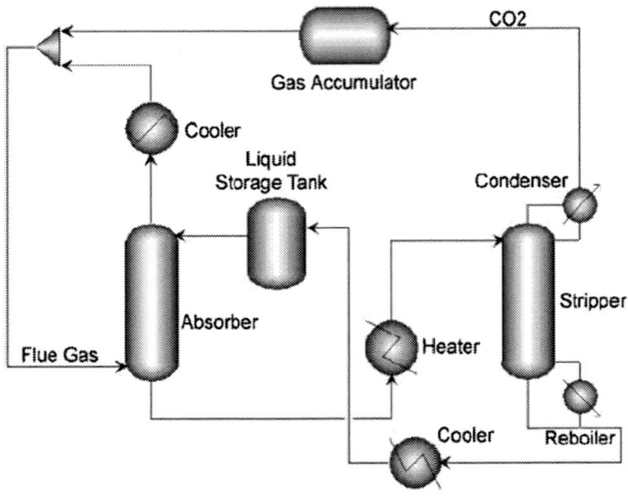

Figure 23. Schematic of CO_2 capture pilot plant from (Dugas, 2006).

A number of studies were also carried out on the process performance in terms of packing performance and absorber/stripper configurations among others. A rate-based model of the absorber unit was developed. Absorber intercooling was found to improve absorption performance especially for high absorption capacity solvents (NETL, 2008).

2.3.3.2. International Test Centre (ITC) for CO_2 Capture

The ITC is formed by the collaboration of University of Regina (Canada) led by Prof. Malcolm Wilson and a consortium of government, industrial partners and its aim is to explore and develop new cost effective technologies for CO_2 capture.

Infrastructure in the ITC consists of bench-scale CO_2 separation units as well as a multi-purpose pilot plant unit (1 ton of CO_2 per day) at the University of Regina. A 250 kW steam boiler is used to generate the flue gas which is then treated in a CO_2 absorption unit. The absorption column is composed of three 0.3 m-diameter sections for a total height of 10 m.

In 2000, the ITC re-commissioned a semi-commercial (4 ton of CO_2 per day) demonstration unit adjacent to SaskPower's 875 MWe Boundary Dam power plant. The unit captures CO_2 from part of the flue gas from this coal-fired power plant (Wilson *et al.*, 2004). This facility consists of three units in series as shown in figure 24:

i. a baghouse unit for fly ash removal.
ii. a scrubbing unit for removal of SO_2 down to 2 ppm.
iii. chemical absorption-based CO_2 recovery unit (Fluor's Econamine FG^{SM} Technology).

Kinetics of the reactive absorption of carbon dioxide in high CO_2 loaded, concentrated aqueous MEA solutions were studied in Aboudheir *et al.* (2003) experiments.

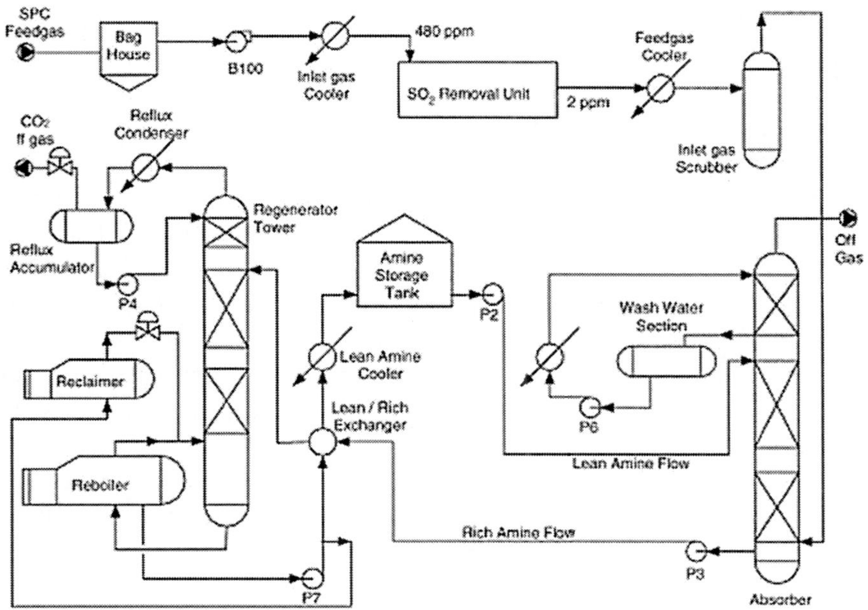

Figure 24. Schematic of boundary dam CO2 pilot plant (Wilson et al., 2004).

Kinetics of the reactive absorption of carbon dioxide with mixed solvents MEA and MDEA were again studied in Edali *et al.* (2009). Pilot plant studies of the CO_2 capture performance of aqueous MEA and mixed MEA/MDEA solvents at the University of Regina CO_2 capture technology development plant and the Boundary Dam CO_2 capture demonstration plant were compared in Idem *et al.* (2006). Uyanga and Idem (2007) studied the degradation of solvent MEA in the presence of SO_2. Kittel *et al.* (2009) studied corrosion of MEA unit for CO_2 capture through pilot plant experiment.

2.3.3.3. CASTOR

This European Commission-funded and IFP-run project involves capturing and providing geological storage for 30% of the emissions released by large industrial facilities around Europe (mainly conventional power stations, i.e. for 10% of Europe's CO_2 emissions). CASTOR, which started in February 2004, was a 4-year programme and counts members from 11 EU countries, including: (a) 16 industrial firms (Dong Energy, Vattenfall, Repsol, Statoil, Gaz de France, Rohoel, Alstom power, RWE, etc.); (b) 12 research institutes (IFP, BGRM, Imperial College, TNO, BGS, etc.). The CASTOR project had a total budget D 15.8 million with a contribution of 8.5 million from European Commission (FP6 – sixth framework programme) (IEA GHG, 2010). Its specific goals involve halving the cost of capturing (from 40–60/ton CO_2 to 20–30/ton CO_2) and separating CO_2, developing the geological-storage concept's efficiency, safety and security while limiting its environmental impact, and testing it in real-life, industrial-scale facilities.

A gas-fired mini plant with full absorption/desorption cycle was built at the University of Stuttgart. This facility consists of an absorber with 0.125 m diameter absorber and 4 m height and a 2.5 m high stripper. An industrial-scale pilot plant facility (figure 18) was launched alongside a power plant run by Dong Energy (formerly ELSAM) in Esbjerg, Denmark on 15 March 2006. This plant has a capacity to capture about 24 ton of CO_2 per day. Studies were carried out on the selection of solvents as well as solvent degradation for the larger CASTOR pilot plant (Notz *et al.*, 2007). Since March 2006, four 1000-h test campaigns have been carried out with MEA solvent and new proprietary solvents CASTOR-1 and CASTOR-2. In January to February 2006, a 1000-h preliminary test campaign was conducted using 30 wt% MEA being the reference solvent. Another 1000-h test was repeated from mid December 2006 to February 2007 to improve on certain problems encountered in the first test while collecting data (Knudsen *et al.*, 2009). Results show that it is possible to run the post combustion plant continuously while achieving roughly 90% CO_2 capture levels. In addition, one of the proprietary solvents, CASTOR-2, operated with lower steam requirement and L/G ratio than the conventional MEA solvent. From CASTOR project, it was concluded that future investigations would involve tests of solvents developed in the EU

CESAR project. In addition, the effect of process modifications on steam requirements as well as environmental effects would be investigated (Knudsen *et al.*, 2009).

2.3.3.4. iCap

iCap is a 4-year project supported by the European Commission under the 7th Framework Programme. It consists of a consortium of 15 partners including 8 Research and Technical Development providers and 6 power companies as well as an Australian Research Institute and a Chinese University. The project started on January 1, 2010.

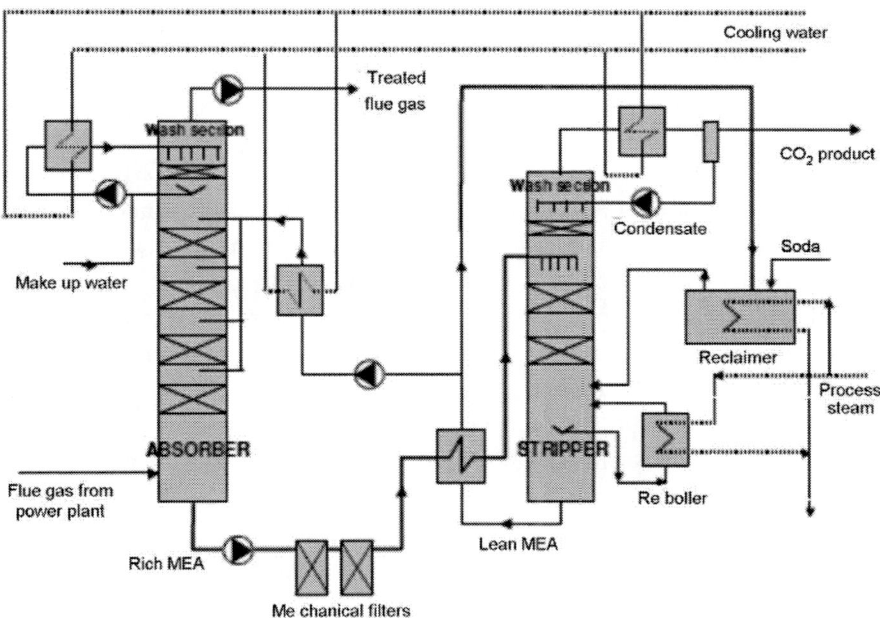

Figure 25. Simplified flow diagram of the CASTOR pilot plant at Esbjergvaerket (Knudsen et al., 2009).

iCap aims to remove barriers that cause bottlenecks in post-combustion and pre-combustion CO_2 capture. Targets include halving the efficiency penalty of CO_2 capture for power plants and reducing the associated CO_2 avoidance cost to 15/ton CO_2. These aims should accelerate the commercial development of large scale near zero emission power generation technology based on CCS (iCap, 2010).

2.3.3.5. CAPRICE

CAPRICE is funded by the European Union and was scheduled to last two years. This project began on 1 January 2007. It was run by TNO, a Dutch organization. CAPRICE stands for CO_2 capture using Amine Process International Cooperation and Exchange and involves pooling information and research findings on amine-enabled CO_2 capture with non-European Carbon Sequestration Leadership Forum (CSLF) countries. More specifically, findings from the European CASTOR project's MEA chapter will be compared with those of the ITC at the University of Regina (Canada). It counts: (a) 10 research centres (University of Regina, Alberta Research Council, ITC, Energy Inet, IFP, Trondheim University, Stuttgart University, Tsinghua University, Topchiev Institute of Petrochemical Synthesis, and Salvador University); (b) 3 power generation companies (E-ON, Dong Energy and Vattenfall) (Caprice, 2010).

2.3.3.6. CESAR

This 4-year long FP7 (Seventh Framework Programme)-funded project was launched in 2008 and aims for a breakthrough in the development of low-cost post-combustion CO_2 capture technology to provide economically feasible solutions for both new power plants and retrofit of existing power plants which are responsible for the majority of all anthropogenic CO_2 emissions. CESAR focuses on post-combustion as it is the only feasible technology for retrofit and current power plant technology. The primary objective is to decrease the cost of capture down to 15/ton CO_2. The consortium consists of 3 research organizations, 3 universities, 1 solvent supplier, 1 membrane producer (SME), 3 equipment suppliers, 2 oil and gas companies and 6 power generators (with industrial commitment) (CO_2cesar, 2010). This project employs the Esbjerg pilot plant used in the CASTOR project. Novel activities and innovations CESAR focuses at are (CO_2cesar, 2010):

- Novel (hybrid) solvent systems.
- New high flux membranes contactors.
- Improved modeling and integration studies at system and plant level.
- Testing of new solvents and plant modifications in the Esbjerg pilot plant. In the Esbjerg pilot plant novel technologies are assessed and compared with mainstream techniques to provide a fast track towards further scale-up and demonstration.

2.3.3.7. The Joint UK–China Near Zero Emissions Coal (NZEC) Initiative

The project is managed by UK consultants AEA Technology plc (for the Department Energy and Climate Change in the UK), in partnership with the Administrative Centre for China's Agenda 21 (ACCA21), and involves a consortium of 28 industrial and academic partners from the UK and China. Its objectives are (a) knowledge sharing and capacity building; (b) future technology perspectives; (c) case studies for carbon dioxide capture; (d) carbon dioxide storage potential; and (e) policy assessment. The project started in 2007 and was finished at the end of 2009 (NZEC, 2009)

2.4. Adsorption

A number of solids can be used to react with CO_2 to form stable compounds at one set of operating conditions and then, at another set of conditions, be regenerated to liberate the absorbed CO_2 and reform the original compound. However, solids are inherently more difficult to work with than liquids, and no solid sorbent system for large scale recovery of CO_2 from flue gas has yet been commercialized, although molecular sieve systems are used to remove impurities from a number of streams, such as in the production of pure H_2 (Figueroa et al., 2008).

2.4.1. Molecular Sieve

Molecular sieves comprise a range of specially designed sieves that are capable to separate molecules based on their molecular weight or molecular size. This technology is believed to be cost-effective and can be adapted to several carbon sequestration schemes (Stewart and Hessami, 2005). There were many research activities aiming to improve the CO_2 adsorption by chemically treating the molecular sieve surface. Adsorbents based on high surface area inorganic supports, incorporating basic organic groups, such as amines or its derivatives, are of particular interest. The interactions between the basic surface and acidic CO_2 molecules are thought to result in the formation of surface ammonium carbamate compounds under anhydrous conditions and also as ammonium bicarbonate and carbonate species in the presence of water. Figure 18 shows, schematically, the reactions between CO_2 and amines within an adsorbent environment. Similar to the amine absorption process, described previously, the theoretical CO_2 adsorption capacity is of 0.5 CO_2/mol surface-bound amine group without the presence of water, which means 1.0 mol CO_2/mol surface-bound amine in the presence of water. Other mesoporous substrates, such as silica (Knowles *et al.*, 2005; Knowles *et al.*, 2006), SBA-1 (Yoshitake *et al.*, 2003), SBA-15 (Gray *et al.*, 2005), MCM-41 (Song, 2006; Xu *et al.*, 2005; Yoshitake *et al.*, 2003) and MCM-48 (Huang *et al.*, 2003), are especially interesting as they possess pores that are large enough to be accessed by molecules incorporating the amino groups. Both the porosity and surface functional groups facilitate the capture of CO_2.

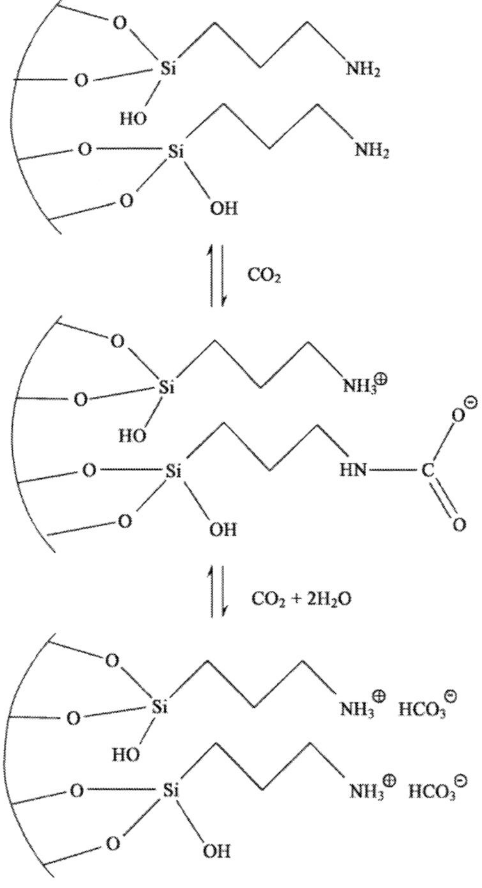

Figure 26. Surface reactions of amine groups with CO_2 (Knowles *et al.*, 2005).

MCM-41

$$\left(-NHCH_2CH_2-\right)_x\left(-\underset{\displaystyle CH_2CH_2NH_2}{\underset{|}{N}}-CH_2CH_2-\right)_y$$

Polyethylenimine

- **Uniform, "tailorable" 15-100 A mesopores**
- **Narrow pore size distribution**
- **High surgace area**
- **Large pore volume**
- **High thermal stability**
- **High amine group density, large amount of CO_2 adsorption sites**

- **Secondary and tertiary amine facilitate desorption**
- **Viscous liquid**
- **Slow CO_2 adsorption/desorpt kinetic**

Figure 27. Structures of MCM-41 and PEI (Song, 2006).

2.4.2. "Molecular Basket" Adsorbent for CO_2 Separation

Song (2006) and Xu *et al.* (2002) studied a novel CO_2 "molecular basket" adsorbent. This adsorbent is based on mesoporous molecular sieve of MCM-41 impregnated with polyethylenimine (PEI), which are shown in Figures 27 and 28 shows the concept of this molecular structure. Along with the increase in PEI loading, the surface area, pore size and pore volume of the loaded MCM-41 decrease. When the loading is higher than 30% (per weight), mesoporous pores began to be filled with PEI, and this sorbent exhibits a synergetic effect on the adsorption of CO_2. At a PEI loading of 50% (per weight), the highest CO2 adsorption capacity of 246 mg/g PEI is obtained, which is 30 times higher than that of MCM-41 and this is about 23 times higher than the one of pure PEI. Chemical impregnation proved to be the best procedure to prepare this molecular basket. The previous study also shows that the molecular basket can selectively capture CO_2 regarding the separation of CO_2 from simulated flue gas, the separation of CO_2 from natural gas-fired and coal-fired boiler flue gas (Song, 2006).

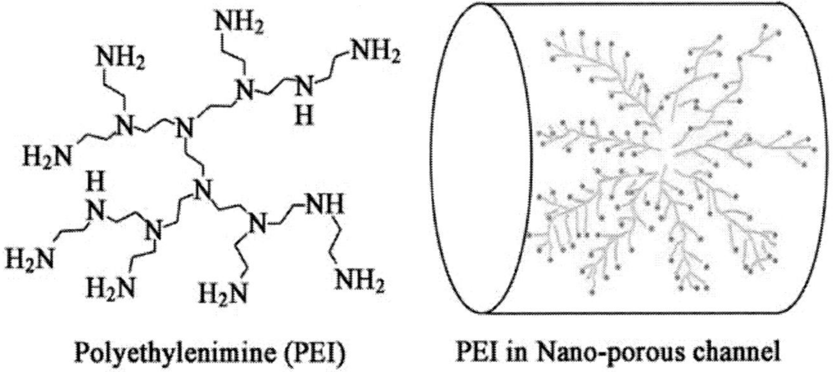

- **Large pore volume nano-porous support can store large amount of CO_2**
- **Branched CO_2-affinity polymers provide adsorption sites**
- **Branched amine facilitate the desorption**
- **Synergic effect on the CO_2 adsorption capacity and adsorption kinetic between nano-porous support and polyethylenimine**

Figure 28. "Molecular basket" concept for highly-selective high-capacity CO2 adsorbent (Song, 2006).

2.4.3. Adsorption by Activated Carbon

Anthracites are known to produce high surface area activated carbon since long. Maroto-Valera *et al.* studied the CO_2 capture behavior of steam-activated anthracite (Maroto-Valera *et al.*, 2005) and they found that CO_2 capture does not show a linear relationship with the surface area. The highest CO_2 adsorption capacity was found to be 65.7 mg CO_2/g adsorbent for the anthracite activated at 800 °C for 2 h, having a surface area of 540 m2/g. However, the anthracite with the highest surface area of 1071 m2/g only had a CO_2 adsorption capacity of 40 mg CO_2/g adsorbent, which can be explained by certain size pores being effective for CO2 adsorption. Also the NH3 treatment and PEI impregnation increased the CO_2 capture capacity of the activated anthracites at higher temperature, due to the introduction of alkaline nitrogen groups on the surface.

Table 4. Adsorption characteristics of some sorbents

	Sorbent	**Adsorption capacity (mg CO_2/g)**	**References**
	NH_3 in water solution	1200	Resnik *et al.*, 2004
MCM	MCM41 modified with PEI	82.0 – 88.0	Xu *et al.*, 2005 ; Soong, 2006
	MCM41 modified with TEPA	211.0	Yue *et al.*, 2008
Silicates	Hexagonal mesoporous silica modified with DT	59.0	Knowles *et al.*, 2005
	Lithium silicate	360.0	Kato *et al.*, 2005
Zeolites	13X	16.0 – 233.2	Chatti *et al.*, 2009 ; Siriwadane *et al.*, 2003
	4A	22.0 – 211.2	Siriwadane *et al.*, 2003
	5A	16.7 – 138.6	Siriwadane *et al.*, 2003
	Y60 modified with TEPA	112.7 – 189.9	Su *et al.*, 2010
Activated charcoal	Anthracite	65.7	Maroto-Valera *et al.*, 2005
	BPL	323.2	Himero *et al.*, 2005
	Monolytes	317.5	Grande *et al.*, 2008
	Norit (R1 and RB2)	419.2 – 454.1	Yang *et al.*, 2001

2.4.4. CO_2 Adsorbents Based on Lithium Compounds

Lithium zirconate (Li_2ZrO_3), having favorable CO_2 sorption characteristics, has also been investigated as a high temperature CO_2 absorbent (Fauth *et al.*, 2005). This technology is based on the chemical reaction using Li_2ZrO_3 to capture CO_2, as follows:

$$Li_2ZrO_3(s) + CO_2(g) \longleftrightarrow Li_2CO_3(s) + ZrO_2(s)$$

This reaction is reversible in the temperature range of 450–590 °C: the reaction can be easily reversed by a simple temperature swing approach, and the formation of eutectic carbonate composed of Li_2CO_3 and K_2CO_3 can accelerate the CO_2 absorption reaction. A number of binary and ternary eutectic salt-modified lithium zirconate sorbents were also identified and evaluated for high temperature CO_2 capture. The combination of binary alkali carbonate, binary alkali/alkaline earth carbonate, ternary alkali carbonate and ternary alkali carbonate/halide eutectic to Li_2ZrO_3 noticeably improved the CO_2 uptake rate and CO_2 sorption capacity. The formation of a eutectic molten carbonate layer on the outer surface of reactant Li_2ZrO_3 particles seem to facilitate the transfer of gaseous CO_2 while the sorption process take place. Lithium silicate (Li_4SiO_4) was studied by Nakagawa *et al.* on their CO_2 adsorption behaviors (Essaki *et al.*, 2004; Kato *et al.*, 2005). They found that the capacity of lithium silicate is much larger than that of lithium zirconate. Lithium silicate adsorbs CO_2 below 720 °C and releases CO_2 above 720°C by the following mechanism:

$$Li_4SiO_4 + CO_2 \longleftrightarrow Li_2SiO_3 + Li_2CO_3$$

Desired features, such as large capacity, rapid absorption, wide range of temperature and CO_2 concentrations, as well as stability, make this compound a strong candidate for developing commercially competitive CO_2 adsorbents. Table 4 compares the adsorption characteristics, in terms of CO_2, for the above mentioned sorbents.

2.4.5. Novel Sorbent Substances

Metal organic frameworks (MOFs) are a new class of hybrid material built from metal ions with well-defined coordination geometry and organic bridging ligands. They are extended structures with carefully sized cavities that can adsorb CO_2. High storage capacity is possible, and the heat required for recovery of the adsorbed CO_2 is low. Over 600 chemically and structurally diverse MOFs have been developed over the past several years. MOF-177 (Willis *et al.*, 2006) has shown one of the highest surface areas and adsorption capacity for CO_2 at elevated pressure. Additional work is still needed to determine stability over thousands of cycles and the effect of impurities at typical flue gas temperature and pressure. UOP has developed a virtual high throughput screening (VHTS) model to reduce the number of MOF synthesis experiments to only those that have the highest probability of meeting the DOE sequestration performance and cost metrics. Because there are an unlimited number of possible MOF structures that can be prepared, the VHTS model, which has a high correlation with laboratory measurements on synthesized MOFs, is a valuable screening tool. A team, made from UOP, the University of Michigan, and Northwestern University, is exploring these materials with the objective of developing a process that can recover CO_2 from the flue gas of a PC-fired power plant. Desirable characteristics of MOFs are low energy requirement for regeneration, good thermal stability, tolerance to contaminants, attrition resistance, and low cost. Figure 29 shows the structure of MOF-177, which has an absorption capacity of 1.474 gCO_2/g (Willis *et al.* 2006).

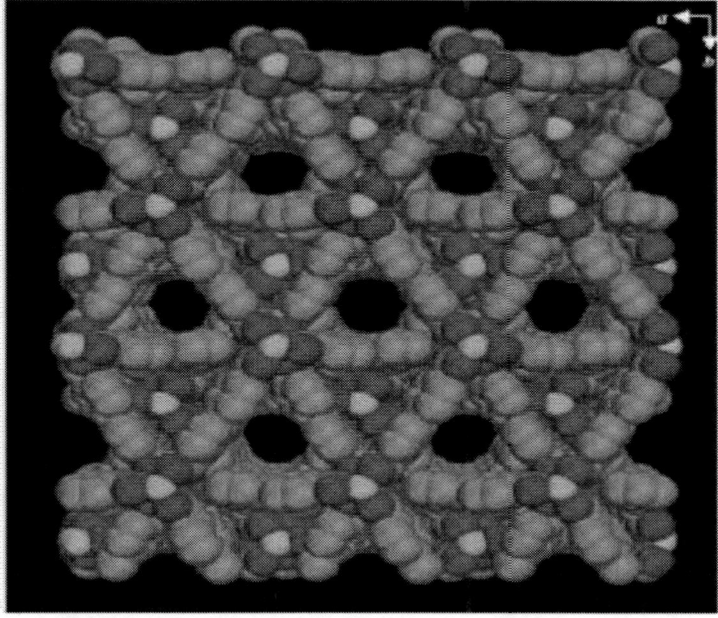

Figure 29. Structure of MOF-177.

Another class of compounds which are capable of absorbing relatively large amounts of CO_2 is Zeolite Imidazolate Frameworks (ZIF), which are porous crystalline materials with tetrahedral networks that resemble those of zeolites: transition metals such as zinc and cobalt replace tetrahedral coordinated atoms, such as silicium, and imidazolate links replace oxygen bridges. An important feature of these materials is that the structure adopted by a given ZIF is determined by link-link interactions, rather than by the structure directing agents used in zeolite synthesis. As a result, systematic variations of linker substituents have yielded many different ZIF that exhibit known or predicted zeolite forms. These materials are chemically and thermally stable, yet have the long-sought-after design flexibility offered by functionalized organic links and a high density of transition metal ions. ZIF selectively capture carbon dioxide from several different gas mixtures at room temperature, with ZIF-100 capable of storing 28 liters of material at standard temperature and pressure (Banerjee et al., 2008). These characteristics, combined with their high thermal and chemical stability and ease of fabrication, make ZIF promising candidate materials for strategies aimed at ameliorating increasing atmospheric carbon dioxide levels (Wang *et al.*, 2008). Figure 30 shows a typical ZIF structure, and figure 31 shows the size of these structures in comparison with other structures. Table 5 compares the adsorption capacity of CO_2 of ZIF to other advanced materials.

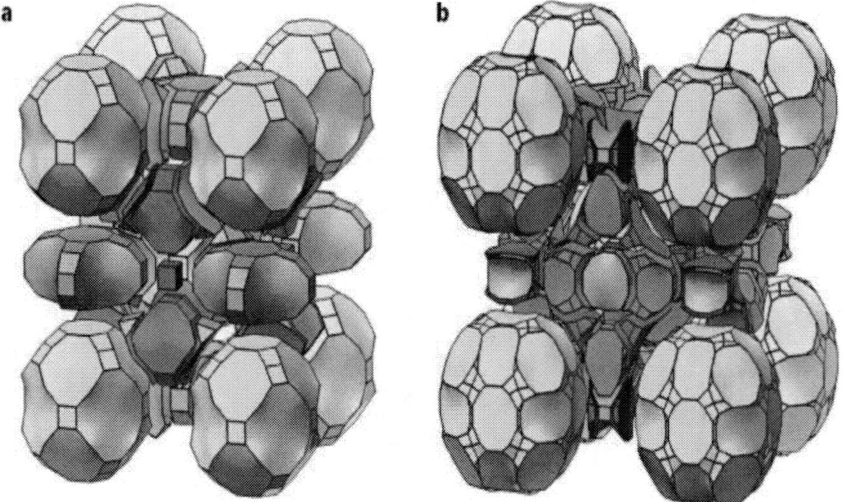

Figure 30. A typical ZIF structure (Banerjee et al., 2008).

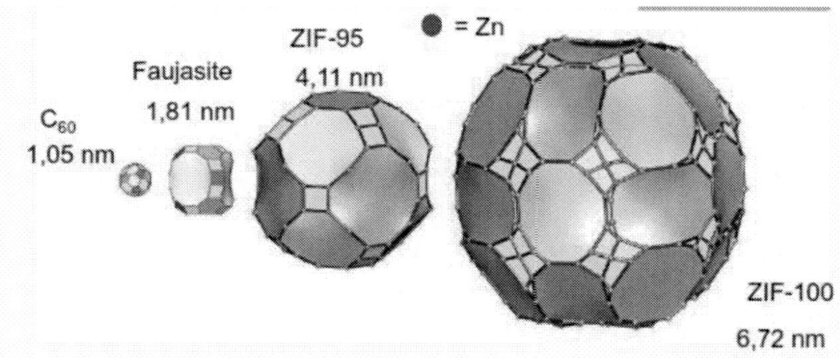

Figure 31. Comparison of the size ZIF structures to other structures (Phan et al., 2010).

2.5. CO_2 Capture Based on Membrane Separation

There are a variety of options for using membranes to recover CO_2 from flue gas. In one concept, flue gas would be passed through a bundle of membrane tubes, while an amine solution flowed through the shell side of the bundle. CO_2 would pass through the membrane and is absorbed by the amine, while impurities would be blocked from the amine, thus decreasing the loss of amine as a result of stable salt formation. Also, it should be possible to achieve a higher loading differential between rich amine and lean amine. After leaving the membrane bundle, the amine would be regenerated before being recycled (Figueroa *et al.*, 2008).

Table 5. CO_2 adsorption capacity of some novel materials

\	**Adsorbent**	**Adsorption Capacity (mg CO_2/g)**	**Refs.**
Carbon nanotubes	Simple wall	87,0	Cinke *et al.*, 2005
	Multiple wall	40,9-96,3	Hsu *et al.*, 2005
	Multiple walls modified with APTS	58,0-114,0	
MOF	MOF 2	140,8	Millward e Yaghi, 2003
	MOF 177	1474	
ZIF	ZIF 69	55,0	Phan *et al.*, 2010
	ZIF 70	74,8	
	ZIF 100	44,0	

Membranes have been widely used in several industrial separations for the last two decades. Industrial applications are currently dominated by polymeric membranes due to the ease of production and design characteristics. Recent research is directed towards the development and, thus, application of inorganic membranes is advancing faster because of the demand in new application fields, such as fuel cells, membrane reactors, and other high-temperature separations. Inorganic or polymeric membrane separation processes are expected to be more effcient than conventional CO_2 separation processes (Damle and Dorchak, 2001; Xu *et al.*, 2001). Both polymeric and metallic membranes can be used to produce clean fuel from a syngas obtained either from coal gasification or by reforming natural gas or methanol. Syngas containing H_2 is delivered at pressures and temperatures which are dependent on the type of fuel processor. However, cooling is required for polymeric membranes. If the fuel gas is composed mainly of CO and H_2, various polymers can be used for hydrogen separation, but the actual selectivity is lower due to polymer plasticization (Carapellucci and Milazzo, 2003). For inorganic membranes, palladium/silver alloys seem to be the most effcient: their 100 percent selectivity for hydrogen guarantees a pure hydrogen permeant. The separation based on metallic membranes is more effcient because the pressure drop across the membrane is smaller. The diffusional characteristics of a particular gas through a given membrane depend on membrane physical and chemical properties, and also on the nature of the permeant species. The third factor, interaction between membrane and permeant, refers to the sorptivity or solubility of the gas in the membrane (Shekhawat *et al.*, 2003). It is important to construct membranes into a practical configuration. Most polymeric membrane gas separation studies are performed on dense homogeneous membranes. Such a configuration provides the permeability with a closer relationship to the intrinsic permeability of the polymer. However, the construction of a gas separation module for practical applications requires a more complex membrane structure: dense membranes tend to be significantly thicker than the selective layer of asymmetric membranes, which leads to dense membranes possessing considerably lower gas fluxes than other alternative membrane structures. Asymmetric composite hollow fiber or flat sheet configuration are often used for this purpose.

2.5.1. Polymeric Membranes

Usually, gas molecules transport through a polymeric membrane takes place by a solution-diffusion mechanism. Other mechanisms include a molecular sieve effect and Knudsen diffusion (Powell and Qiao, 2006). These transport mechanisms are briefly described in section 2.4.2. Terms such as permeability and selectivity are used to describe the performance of a gas separation membrane. There appears to be a trade-off between selectivity and permeability. Gas molecules tend to move through free volumes–the gaps between polymeric structures. Because of the movement of the polymer chains, a channel between gaps can be formed allowing gas molecules to move from one gap to another and thus gas molecules can effectively diffuse through the membrane structure. Selective transport of gases can be achieved by use of a polymer which forms channels of a certain size. Naturally, large channels will allow faster diffusion of gases through a membrane at the cost of less selectivity. Membranes are a low cost means to separate gases when high purity is not critical. However, there is a number of issues associated with the capture of carbon dioxide from flue gas which limit the use of membranes: the concentration of carbon dioxide in flue gases is somewhat low, which means that large quantities of gases will have to be processed. The high temperature of flue gases will rapidly destroy a membrane, so the gases need to be cooled to below 100 °C prior to membrane separation. The membranes also need to be chemically resistant to the harsh chemicals contained within flue gases, or these compounds will have to be removed prior to the membrane separation. Additionally, creating a pressure difference across the membrane will require significant amounts of power. Polymers studied in various studies include: polyacetylenes (Stern, 1994), polyaniline (Illing *et al.*, 2001), poly(arylene ether)s (Xu *et al.*, 2002), polyarylates (Pixton and Paul, 1995), polycarbonates (Aguilar-Vega and Paul, 1993), polyetherimides (Li and Freeman, 1997), poly(ethylene oxide) (Lin and Freeman, 2004), polyimides (Stern *et al.*, 1989), poly(phenylene ether) (Aguilar-Vega and Paul, 1993), poly(pyrrolone)s (Zimmerman and Koros, 1999) and polysulfones (Aitken *et al.*, 1992). Figure 32 shows molecular structures of some commonly used polymers. The performance of some polymeric membranes are summarized in Table 6 mainly separating post-combustion flue gas with CO_2 /N_2 being the main components (Powell and Qiao, 2006). Single-stage membrane systems are not capable of high capture effciency and CO_2 can be further concentrated by a second membrane stage. Materials for effective separation of gases can follow one of two overall strategies: increasing the rate of diffusion of carbon dioxide through the polymeric structure and increasing the solubility of carbon dioxide in the membrane. The introduction of mixed-matrix membranes may allow higher performance which combines the advantages of polymeric and inorganic membranes materials.

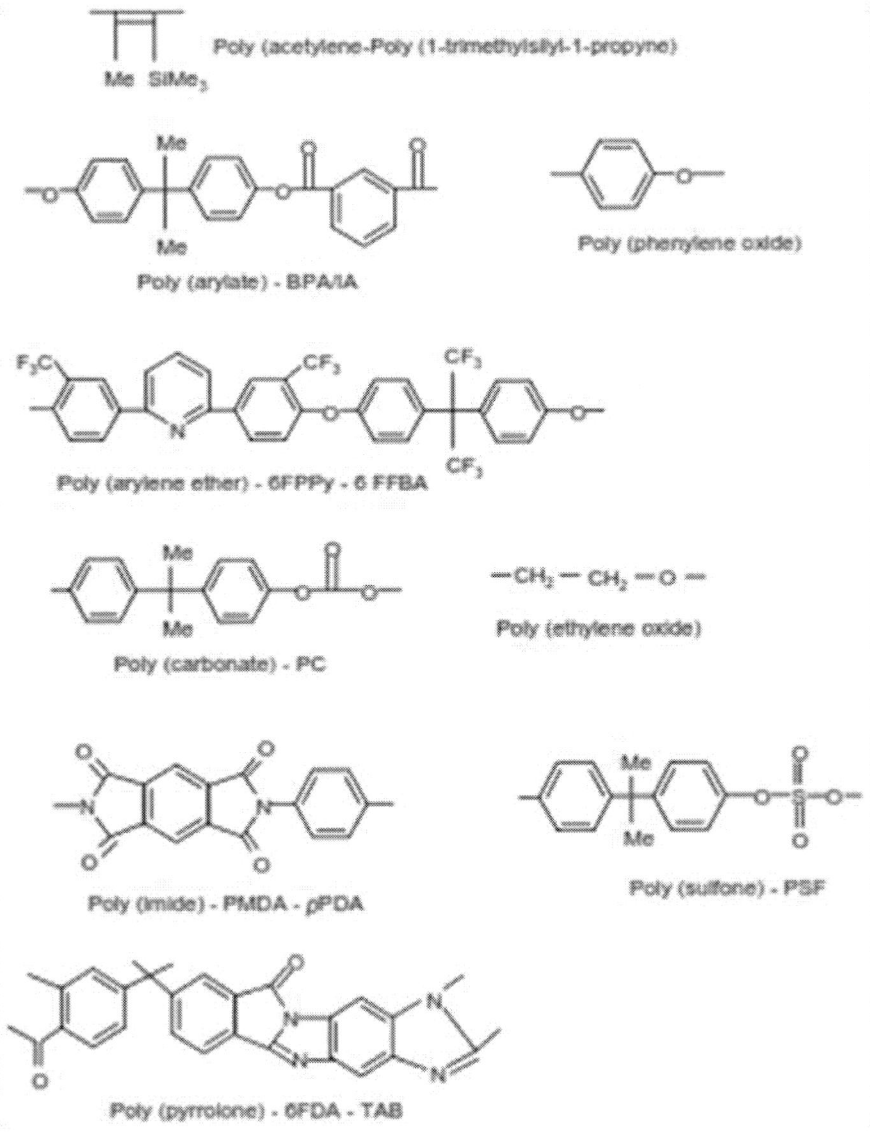

Figure 32. Molecular structures of polymers used in membranes for CO_2 capture (Olajire, 2010).

Table 6. Performance of membranes for separating CO_2 from N_2 (Powel and Qiao, 2006)

Material	Permeance to CO_2 ($m^3 m^{-2} Pa^{-1} s^{-1}$)	Selectivity CO_2 / N_2
Poliimide	735	43
Polydimethylphenylene oxyde	2750	19
Polyssulfone	450	31
Polyetherssulfoea	665	24,7
Poly(4-vinylpiridine) /Polyeterimide	52,5	20
Polyacrilonitrile with polyethyleneglycol	91	27,9
Poly(amide-6b-ethylene oxide)	608	61

2.5.2. Inorganic Membranes

Inorganic membranes can be classified into two categories based on its distinctive structure: porous and dense. In porous inorganic membranes, a porous thin top layer is casted on a porous metal or ceramic support, which provides mechanical strength but offers minimum mass-transfer resistance. Alumina, carbon, glass, silicon carbide, titania, zeolite, and zirconia membranes are mainly used as porous inorganic membranes supported on different substrates, such as α-alumina, γ-alumina, zirconia, zeolite, or porous stainless steel. Surface modification by covalent bonding of a layer of selected compounds with appropriate functional groups is one of the easiest ways to alter membrane performance. Such modification can increase the performances by changing the mean pore size and promoting an eventual specific interaction between the surface of the membrane and the permeating molecules in order to enhance permeation. The dense inorganic membranes consist of a thin layer of metal, such as palladium and its alloys, or solid electrolytes, such as zirconia. The obtained membranes are highly selective for hydrogen or oxygen separation. Gas transport occurs via solution-diffusion mechanism or charged particles in dense membranes. The low permeability across the dense inorganic membranes limits its wide applications as compared to porous inorganic membranes. There are four main transport mechanisms by which gas separation using porous inorganic membranes can be described as shown in Figure 33, and these are: A) Knudsen diffusion, B) surface diffusion, C) capillary condensation, and D) molecular sieving (Shekhawat *et al.*, 2003; Luebke *et al.*, 2006).

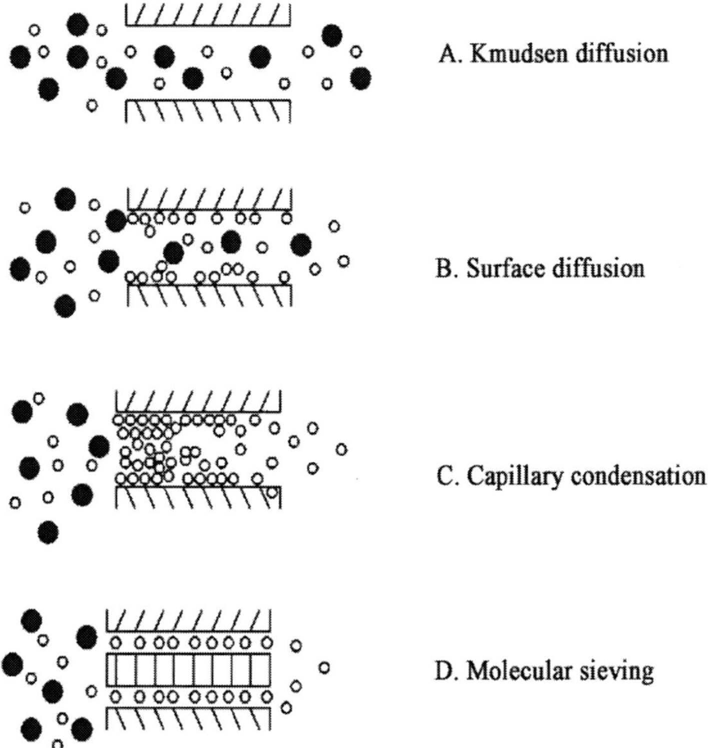

Figure 33. Transport mechanism through microporous membranes (Shekhawat *et al.*, 2003).

2.5.3. Carbon Membranes

Carbon membranes for gas separations are typically produced by the pyrolysis of thermosetting polymers. The pyrolysis temperature, typically in the range of 500 to 1000 °C, depends upon the type of precursor material and defines the separation performance of the carbon membranes (Shekhawat *et al.*, 2003). Pyrolysis of polymeric compounds can be used to form a carbon material with a narrow pore size distribution below molecular dimensions (< 1 nm), which makes it possible to separate gases with very similar molecular sizes. The predominant transport mechanism of these carbon membranes is molecular sieving. The selection of precursor polymers, the membrane preparation method, and the carbonization process determine the performance of carbon molecular sieve membranes. The mechanical stability can be improved by supporting a thin carbon membrane on a porous support material, such as α- alumina. The high thermal and chemical stability of these membranes provide good alternatives for gas separation applications, such as separation of CO_2 in flue gas emissions from power plants.

2.5.4. Alumina Membranes

The generally mesoporous structure of alumina dictates that transport, within these membranes, occurs by a Knudsen diffusion mechanism (Shekhawat *et al.*, 2003). With mixtures such as CO_2/N_2, where the gases have similar mass, and CO_2/H_2, where selectivity towards the heavier component is required, alumina is undesirable as a material for membrane separation layer. Alumina finds its use in the separation of gases mainly as support, where its structural properties, as well as chemical and hydrothermal stabilities beyond 1000 °C make it very useful. Some attempts have been made to modify alumina membranes to facilitate CO_2 surface diffusion with limited success. In order to achieve high separation factors in systems such as CO_2/N_2, an interaction between one of the gases in the mixture and the membrane surface can be introduced by chemically modifying the separation layers.

2.5.5. Silica Membranes

Silica is a viable starting material for the fabrication of CO_2 selective membranes, primarily because of its innate stability and easy modification of its structures. Unlike alumina, which tends to undergo phase transition at relatively low temperatures, or carbon, and can exhibit substantial changes in pore size in oxidizing environments, silica shows exceptional thermal, chemical, and structural stability in both oxidizing and reducing environment. Sol-gel method and chemical vapor deposition (CVD) technique are usually used to prepare silica membranes.

2.5.6. Zeolite Membranes

Zeolites are crystalline aluminosilicates having a uniform pore structure and a minimum channel diameter range of 0.3 to 1.0 nm. The presence of molecular-sized cavities and pores make the zeolites effective as shape-selective materials for a wide range of separation applications. This ability to selectively adsorb molecules by size and polarity is the key to the unusual effciency of synthetic zeolites as the basis for gas separation. Separation occurs in zeolite membranes by both molecular sieving and surface diffusion mechanisms. Zeolite membranes are usually prepared by *in-situ* hydrothermal synthesis on porous stainless steel, α-alumina, or γ-alumina support tubes or disks for the gas permeation studies. Membranes of various zeolites, such as ZSM-5 (Kusakabe *et al.*, 1996), Y type (Kusakabe *et al.*, 1997), silicalite (Lin *et al.*, 2000), A type (Aoki *et al.*, 2000), P type (Dong and Lin, 1998), modernite (Bernal *et al.*, 2000), and silicoaluminophosphate (Poshusta *et al.*, 2000) have been synthesized on porous supports of different nature. Typically, the heat of adsorption of gases on most zeolites increase in this order: $H_2 < CH_4 < N_2 < CO_2$ (Poshusta *et al.*, 2000), which is consistent with the electrostatic properties of each molecule. Carbon dioxide preferentially permeates in CO_2/N_2, CO_2/CH_4, and CO_2/H_2 mixtures at low temperatures, as CO_2 adsorbs more strongly on zeolites than the other gases. For the CO_2/N_2 and CO_2/CH_4 mixtures, CO_2 is smaller in size and thus permeates faster at elevated temperatures. However, selectivities are the same for the CO_2/N_2 or CO_2/CH_4 systems at elevated temperatures due to the absence of competitive adsorption.

2.5.7. Mixed-Matrix and Hybrid Membranes

These molecular sieves possess superior gas transport properties, but have significant problems with their processibility. Incorporation of molecular sieves within a polymer membrane possibly provides both the processibility of polymers and selectivity of molecular sieves (Nomura *et al.*, 1997). Many research is directed to mixed-matrix membranes; the polymer-zeolite pairs include polydimethylsiloxane-silicalite (Tantekin-Ersolmaz *et al.*, 2000), polyimide-carbon molecular sieve (Vu *et al.,* 2003), polyimide-silica (Kusakabe *et al.*, 1996), Nafion-zirconium oxide (Apichatachutapan *et al.*, 1996), HSSZ-13-polyetherimide (Husain and Koros, 2007), acrylonitrile butadiene styrene-activated carbon (Anson *et al.*, 2004). It is known that the permeability of a gas through a zeolite-filled polymeric membrane depends on the intrinsic properties of the zeolite and the polymer (Tantekin-Ersolmaz *et al.*, 2000). However, the performance seems to suffer from defects caused by poor contact at the molecular sieves/polymer interface. This can allow the gases to flow non-selectively around the solid particles. Sonication and decantation techniques have been suggested to overcome this issue (Vu *et al.*, 2003). The successful implementation of this membrane development depends on both, the selection of polymeric matrix and inorganic zeolite, and the elimination of interfacial defects (Anson *et al.*, 2004). A similar concept takes place with hybrid membranes where a porous inorganic support material is modified at the surface with chemicals which have good affinity with CO_2. Strictly speaking, the hybrid membrane might better be named surface-modified inorganic membrane, in order not to be confused with mixed-matrix membrane. This helps CO_2 separation in two ways: porous inorganic materials allows for large fluxes while the chemicals provide selectivity (Luebke *et al.*, 2006; Shekhawat *et al.*, 2003). The studied chemical-inorganic pairs include polyether-silica (Kim *et al.*, 2005), trichlorosilane-γ-alumina (Luebke *et al.*, 2006), organosilane-Vycor glass (Singh *et al.*, 2004), tetrapropylammonium-silica (Yang *et al.*, 2002), titania-trimethoxysilaen (Abidi *et al.*, 2006), γ-alumina- trimethoxysilane (Abidi *et al.*, 2006), hexagonal mesoporous silica-aminopropylhydroxysilyl (Knowles *et al.*, 2005; Chaffee, 2005).

2.5.8. Facilitated Transport Membranes

Facilitated transport membranes (FTM) have received a lot of attention in gas separations because they are offering higher selectivity and larger fluxes (Shekhawat *et al.*, 2003). Higher selectivity in FTM is achieved by incorporating a carrier agent into the membrane, which reacts reversibly with the penetrating species. In addition to the solution-diffusion mechanism of the polymeric membranes, FTMs also involve a reversible complexing reaction. The permeating species dissolves in the upstream portion of the membrane and reacts with the carrier agent inside the membrane to form a complex. This complex diffuses across the membrane, and then releases the permeants on the downstream side of the membrane, while the carrier agent is simultaneously recovered and diffuses back to the feed side. FTMs can be in a form of fixed carrier membranes, solvent-swollen polymer membrane, and mobile carrier membranes (Shekhawat *et al.*, 2003).

2.5.9. Membranes in Conjunction with Chemical Absorption

The combination of chemical absorption and selective membranes has also been suggested, so as to perform absorption and desorption in a single unit (Mano *et al.*, 2003; Okabe *et al.*, 2003). Guha *et al.* (1990) modeled and measured the permeabilities and separation factors through a liquid membrane for a CO_2/N_2 system over a wide range of CO_2 partial pressures. In their approach, an immobilized liquid membrane (ILM, also referred to as supported liquid membrane or SLM) and an aqueous solution of diethanolamine (DEA) was used. In this system, DEA was immobilized in the pores of hydrophobic microporous polypropylene membrane, and helium gas was used as the sweep. Their combined CO_2 absorption and desorption into a single unit that require no external energy was reported to achieve CO_2/N_2 separation factors of 230/516. The model developed by Guha et al. was adopted by Bao and Trachtenberg (2005) to evaluate the performance of hollow fiber, contained liquid membrane (HFCLM) permeator for the separation of CO_2 from a CO_2 and air mixture, using a DEA solution as the liquid membrane by means of both experimental and numerical methods. A permeance of 1.51×10^8 mol/m^{2}sPa was reached and CO_2/N_2 selectivity of 115, with a 20% (wt) DEA liquid membrane and a feed of 15% CO_2 in CO_2/air mixture at atmospheric pressure. Their model predictions compared well with the experimental results at CO_2 concentrations of industrial importance.

2.6. Ionic Liquids

Ionic liquids (ILs) is a broad category of salts, typically containing an organic cation and either an inorganic or organic anion (figure 34) shows the computed electron density for a CO_2 molecule interacting with the ionic liquid [hmim][Tf2N]. The cation [hmim], charge +1, is shown along the top. The anion [Tf2N], charge 1, is along the bottom. A single CO_2 molecule is shown in between the two. This image is from a quantum mechanical calculation that shows the electron density distribution and indicates how CO_2 interacts with the ionic liquid. The blue regions on the surface show areas of relatively large positive charge, while red areas show large negative charge. Green areas are more or less neutral. ILs can dissolve gaseous CO_2 and are stable at temperatures up to several hundred degrees centigrade.

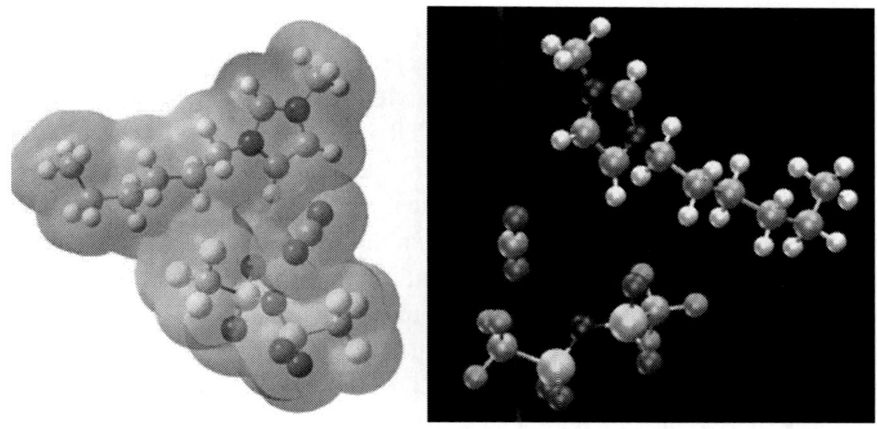

Figure 34. Schematic of an ionic liquid interaction with carbon dioxide (Figueroa, 2008).

Their good temperature stability offers the possibility of recovering CO_2 from flue gas without having to cool it first. Also, since ILs are physical solvents, little heat is required for regeneration. Research at the University of Notre Dame has indicated that, for flue gas application, ILs have demonstrated SO_2 solubility 8 to 25 times that of CO_2 at the same partial pressure (Anderson et al., 2006), thereby allowing this novel solvent to not only remove CO_2 but also serve as an SO_2 polishing step. Collaborative research with NETL scientists has shown that ILs can be used as the separating media for pre-combustion application in supported liquid membranes to separate CO_2 from H_2. Some ionic liquids are commercially available, but the ones most suited for CO_2 separation have only been synthesized in small quantities in academic laboratories. As such, current unit costs are high, but should be significantly lower when produced on a commercial scale for the volumes that would be needed by the power generation sector. The viscosity of many ILs is relatively high when compared to conventional solvents. Viscosities for a variety of ILs are reported to range from 66 to 1110 cP at 20 to 25 °C (Kanel, 2003), and high viscosity may be an issue in practical applications. Based on the finding that the anion is critical in determining CO_2 solubility, several ionic liquids have been developed that have exhibited CO_2 solubilities 40 times greater than achieved prior to the start of the NETL-sponsored research project. Capacity still needs to be significantly improved, however, to meet cost targets. Task specific ILs (TSIL) (Maginn, 2007) that contains amine functionality are being investigated to provide the next step change improvement in CO_2 solubility.

In fact, considerable research work is being done showing high carbon dioxide solubility in certain RTILs, especially in those having imidazolium-based cations. Depending on their selectivity, ILs are stronger candidates for CO_2 capture (Anderson *et al.,* 2007). RTILs portray a typical behavior of a physical solvent (Kamp *et al.,* 2003). The solubility of carbon dioxide, ethylene, ethane, methane, argon, oxygen, carbon monoxide, hydrogen and nitrogen in 1-n-butyl-3-methylimidazolium hexafluorophosphate, [bmim][PF_6] in the temperature range between 10 and 50 °C and pressures up to 13 bar proves superiority of IL over various organic solvents like heptane, cyclohexane, benzene, ethanol and acetone. Dissolution enthalpy and entropy values suggest stronger interaction of CO_2 with the IL. The relatively higher solubility of CO_2 may be attributed to its quadrapole moment and dispersion forces. Owing to their negligible volatility and thermal stability under the explicit conditions, ILs are unlikely to contaminate the gas stream to be cleaned up on top of the RTILs greater thermal stability. Mass transfer of the gas is of much importance especially where gas is to undergo a chemical interaction. Hence, during fabrication of an appropriate ionic liquid, drawbacks posed by high viscosity must be addressed (Anthony et al., 2002). Experimental and simulation studies have shown that CO_2 is more soluble in alkylimidazolium-based ILs. The origin of this high solubility could be related more to the anion moiety that enhances interactions by favoring particular distributions of CO_2 molecules around the cation (Cadena *et al.,* 2004). Raeissi and Peters verified the thermal stability of 1-n-butyl-3-methylimidazolium bis[trifluoromethylsulfonyl]imide,[bmim][Tf_2N], by conducting the gas capture experiments in the temperature range of 40–177 °C and pressures up to 140 bar. Even after keeping at 177 °C for more than 10 h, the ionic-liquid conferred reproducible results for CO_2 solubility (Raeissi and Peters, 2009). Alkyl-side chain length of the imidazolium-based cation of the ILs also affects CO_2 solubility to a certain extent; how- ever, the outcome is not as prominent as that by switching of anion. Fluorine substituted side chains greatly augment the uptake of CO_2 with respect to the corresponding non-substituted side chains but at the expense of an increase in viscosity (Sin and Lee, 2008). The nature of anion seems to have stronger influence on gas solubility than that of the cation. Ionic liquids possessing [Tf_2 N] anion show higher CO_2 solubility among imidazolium-based RTILs. A number of factors like free volume, size of the counter ions, and strength of cation–anion interactions within the ionic-liquid structure seem to govern CO_2 solubility in RTILs. Higher gas solubility with increase in alkyl-side chain may be the result of increased free volume available for CO_2 with corresponding decrease in cation–anion interactions. The thermal stability and negligible volatility make RTILs quite imposing. Hou and Baltus (2007) found that even after regenerating the ionic-liquid six times, by purging N_2 followed by evacuation at 70 °C, there was practically no change in the gas capture capacities. The equilibrium pressure not only depends on temperature but also on CO_2 concentration. At 60 bar, CO_2 solubility in 1-ethyl-3-methylimidazoliumbis[trifluoromethylsulfonyl]imide, [emim][Tf_2N], is found to be 60 mol% which proves the higher efficiency of IL for CO_2 capture. When compared with 1-ethyl-3-methylimidazolium hexafluorophosphate, [emim][PF_6], the gas is found more soluble in IL with [Tf_2N]$^-$ anion. The difference is further pronounced at higher CO_2 mole fraction. Such data is confirming the effect of anion on CO_2 interaction with IL. Fluoroalkyl group enhances the CO_2 solubility, thus making [emim][Tf_2] more efficient for CO_2 capture. Room-temperature ionic liquids can be effectively used for hydrogen purification with high selectivity for CO_2/H_2 separation. Selectivity of the ionic liquid, [bmim][PF_6], for O_2/H_2 mixtures constituting 45–50 wt% H_2 is in the range of 30–300. Selectivity drops at higher temperature but enhances with pressure increase. Hence this setup may be employed in CO_2 capture from pre-combustion power plants. A pressure-swing adsorption/desorption method can be employed for H_2 purification by RTILs. CO_2 showed maximum solubility in 1-ethyl-3-methylimidazolium 2-(2-methoxyethoxy)ethylsulfate, [emim][MDEGSO4] at 30 °C in the pressure range of 8.54–67 bar, and expectably increasing with pressure rise. Pyrrolidinium and ammonium based RTILs like 1-n-butyl-1-methylpyrrolidiniumbis(trifluoromethylsulfonyl) amide([bmpy][Tf_2N]) and trimethyl(butyl)ammoniumbis(trifluoromethyl) sulfonyl)imide([N(4)111][Tf_2N-]) have also been investigated for H_2 purification showing CO_2 absorption capacity comparable to imidazolium-based RTILs in the temperature range of 20–140 ℃ (Kumelan *et al*, 2010). Regarding H_2S/CO_2 selectivity, H_2S was found almost three times more soluble than CO_2 in 1-(2-hydroxyethyl)-3-methylimidazoliumtetrafluoroborate than gases, with a higher partial pressure of CO_2 diminishes this advantage. This observance illustrates that RTILs can be efficiently tailored to remove H_2S and CO_2 concurrently (Heintz, *et al.*, 2009).

To cope with the viscosity constraints, RTILs may be mixed with some common organic solvents or water. Addition of water (IL aqueous solutions) helped to overcome viscosity problems during processing accomplished at the expense of decrease in gas capture ability. This is evident from the behavior of an ionic liquid [Choline][Pro] examined in pure form as well as after mixing with polyethylene glycol (PEG 200) at temperatures 35–80 ℃ and ambient pressure (Li *et al.*, 2008). Gas solubility decreased with increasing amount of PEG 200, under constant temperature and pressure conditions. This is explicable because of the low CO_2 solubility in PEG 200. However, to enhance the rate of both absorption and desorption, addition of an appropriate amount of PEG 200 has been found favorable. This may be due to decrease in viscosity and/or solvent role of PEG 200. Another, more workable, option may be the replacement of aqueous media of alkanolamine systems with some stable and non-volatile room-temperature ionic liquids in order to combine the advantages of both, i.e., negligible vapor pressure, higher thermal stability and lower heat capacity of ionic liquids, and fast capture kinetics and low viscosity of certain alkanolamines. Switching the CO_2 capture product (carbamate in this case) into a foreign phase would pull the equilibrium-limited CO_2 absorption towards higher CO_2 conversion values unlike in conventional aqueous amine solutions with soluble carbamate salt. Thus, it can be inferred that to take advantage of useful properties of ILs, amine-IL solutions need to be investigated more deeply as potential replacement solvents for aqueous amine scrubbing systems.

Regarding natural gas purification, certain hygroscopic imidazolium-based ionic liquids like [bmim][PF_6], [C_8mim][BF_4] and [C_8mim][PF_6] have ability of dehydration as well. Also, the presence of water, along with acetate ion, in some ionic liquids akin to [hmim][acetate] and [bmim][acetate] may facilitate the capture phenomenon through weak bonding with CO_2 (Chime *et al.*, 2006). Diminished corrosion of the equipment, almost one-third the heat capacity of (especially imidazolium-based) RTILs compared to aqueous systems may have profound effect in extenuating the high price outcome. In short, room-temperature ionic liquids especially imidazolium-based RTILs may be employed in natural gas/hydrogen purification or in CO_2 capture from fossil fuel based power plants. Regarding regeneration, room-temperature ionic liquid based materials may be easily recovered either by pressure sweep process coupled with vacuum treatment, by applying heat or by bubbling nitrogen through the absorbent. However, task-specific ionic liquids or RTILs mixed with amine bearing species require temperature sweep regeneration involving vacuum heating. As discussed earlier, CO_2 is sufficiently soluble in room- temperature ionic liquids (RTILs). However, the CO_2 capture ability can be enhanced by introducing basic character in the ILs. Functionalization of ionic liquids with a suitable moiety (like amine) may be implied in this regard. CO_2 absorption ability of TSILs can reach up to threefold of the corresponding RTILs. The enhanced effect of pressure in case of TSILs was observed by the fact that there was a steady increase in gas load with rise in pressure, providing evidence both for chemical as well as physical sorption. The effect is not so apparent in case of aqueous amine solutions which possess stoichiometric limitations. Reversible sequestration of CO_2 is achieved by attaching primary amine moiety to an imidazolium cation, without any dwindle in the ionic-liquid stability. For five consecutive cycles of gas absorption, the regenerated TSIL ([pabim][BF_4]) did not show any loss of efficiency. [pabim][BF_4] exhibits better capture of CO_2 compared to [hmim][PF_6], owing to chemical capture phenomenon in the former. The TSIL when exposed to CO_2 for 3 h at room temperature and pressure, the mass gain was 7.4% which corresponds to 0.5 molar uptake of CO_2 (maximum theoretical value for CO_2 capture as amine carbamate). The TSIL remains stable even after five gas sorption/desorption cycles without any detectable loss in efficiency. The inclusion of water in the ionic liquid was found to increase the CO_2 holding capacity which might be due to the formation of bicarbonate species as well (Bates *et al.*, 2002). In spite of the tunable approach towards TSILs, these functionalized species exhibit much higher viscosities as compared to the corresponding RTILs or other commercially available CO_2 scrubbing solutions, posing too serious complications to be applicable on industrial scale. CO_2 capture results in sharp increase of viscosity of the TSILs, resulting into a gel-like material (Gutowski and Maginn, 2008). This drawback may be avoided by utilizing mixtures of TSILs and RTILs or TSILs may be adsorbed onto porous membranes. Comparison of CO_2 capture by ionic liquids with that by conventional aqueous amine solutions (30 wt% MEA/MDEA) illustrates that the absorption activities of ionic liquids resembles to that of common physical solvents. Nonetheless, CO_2 absorption ability increases significantly on functionalization of ionic liquid with primary amine moiety. Task-specific ionic liquids, [Am-im][BF_4] and [Am-im][DCA], perform like chemical solvents at low pressures ($\leq$1 bar), however, at higher pressures they pursue the performance of room-temperature ionic liquid, [bmim][BF4]. On the other hand, aqueous amine solutions accomplished the maximum capacity at about 2 bar and any further increase in pressure does not seem feasible. Whereas functionalized ionic liquids (TSILs) carry on steady CO_2 absorption with ascending pressure (Shu and Li, 1992). This behavior depicts that TSILs possess both chemical as well as physical tools for gas capture.

Another interesting approach (Luis *et al.*, 2009) is the use of supported ionic liquid membranes: the permeability of air is one order of magnitude lower than CO_2 permeability and it is also lower than the permeability of mixtures of air and 10% volume SO_2.

2.7. Enzime Based Systems

Biologically based capture systems are another potential avenue for improvement in CO_2 capture technology. These systems are based upon naturally occurring reactions of CO_2 in living organisms. One of these possibilities is the use of enzymes. An enzyme-based system, which achieves CO_2 capture and release by mimicking the mechanism of the mammalian respiratory system, is under development by Carbozyme, as shown in figure 35. The process, utilizing carbonic anhydrase (CA) in a hollow fiber contained liquid membrane, has demonstrated at laboratory-scale the potential for 90% CO_2 capture followed by regeneration at ambient conditions. This is a significant technical improvement over the MEA temperature swing absorption process. The CA process has been shown to have a very low heat of absorption that reduces the energy penalty typically associated with absorption processes.

The rate of CO_2 dissolution in water is limited by the rate of aqueous CO_2 hydration, and the CO_2-carrying capacity is limited by buffering capacity. Adding the enzyme CA to the solution speeds up the rate of carbonic acid formation; CA has the ability to catalyze the hydration of 600,000 molecules of carbon dioxide per molecule of CA per second compared to a theoretical maximum rate of 1,400,000 (Trachtenberg et al., 1999). This fast turnover rate minimizes the amount of enzyme required. Coupled with a low make-up rate, due to a potential CA life of 6 months based on laboratory testing, this biomimetic membrane approach has the potential for a step change improvement in performance and cost for large scale CO_2 capture in the power sector. Although the reported laboratory and economic results may be optimistic, the ''Carbozyme biomimetic process can afford a 17-fold increase in membrane area or a 17 times lower permeance value and still be competitive in cost with MEA technology'' (Yang and Ciferno, 2006). The idea behind this process is to use immobilized enzyme at the gas/liquid interface to increase the mass transfer and separation of CO_2 from flue gas. Technical challenges exist before this technology can be pilot tested in the field. These limitations include membrane boundary layers, pore wetting, surface fouling (Boa and Trachtenberg, 2006), loss of enzyme activity, long-term operation, and scale-up, which are being addressed in a current project.

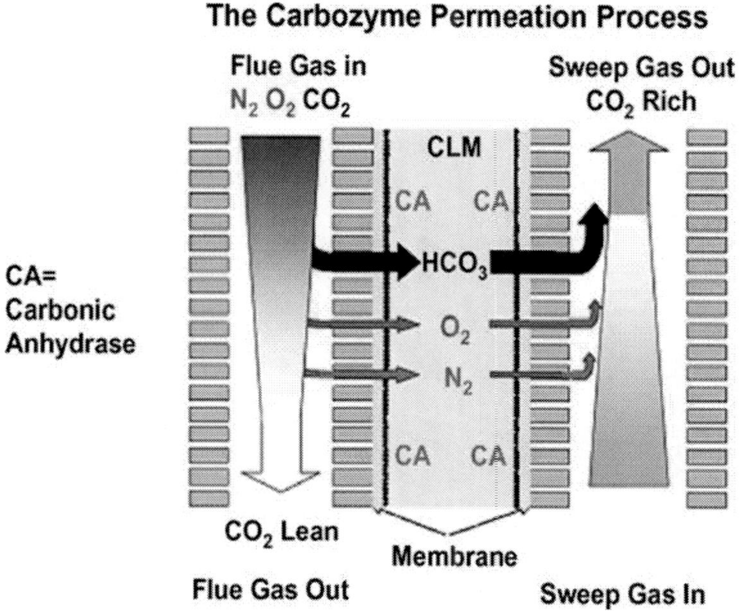

Figure 35. Schematic of the Carbozyme permeation process (Figueroa et al., 2008)

Chapter 3

SOME ALTERNATIVES FOR CO_2 REUTILIZATION

Apart from the geologic storage of CO_2, which was already mentioned in the previous sections, the captured CO_2 is a major source of carbon and, thus, can potentially be used for obtention of other carbon compounds or even hydrocarbons. Some of the reactions involved are given below:

i) Methane synthesis:

$CO_2 + 4H_2 \rightarrow CH_4 + 2H_2O$; $\Delta G^o = 113.6$ kJ/mol

ii) Methanol synthesis:

$CO_2 + 3H_2 \rightarrow CH_3OH + H_2O$; $\Delta G^o = 3.9$ kJ/mol

iii) Ethylene synthesis:

$2CO_2 + 2CH_4 \rightarrow 2CO + C_2H_4 + 2H_2O$ $\Delta G^o = 35$ kJ/mol

iv) Carbohydrates (photosynthesis):

$nCO_2 + nH_2O \rightarrow C_nH_{2n}O_n + nO_2$

As shown, these reactions are endothermic, which mean that large amounts of energy are needed so that these reactions can take place. Apart from that, some of these reactions require hydrogen as reagent. Therefore, some of thes processes are quite difficult to put into practice, thus requiring considerable research and development in order to achieve suitable catalysts to ameliorate and optimize some of the synthesis processes. Nowadays, CO_2 is already being used for synthesis of methanol, which is used as fuel and also as raw material for producing formaldehyde resins; for obtaining low molecular weight parafins and olefins and dimethylcarbonate, which is a gasoline additive (gasoline booster). Low molecular weight olefins are being used for obtaining polycarbonates and also polyethylene.

Potential uses lie within the production of precipitated calcium carbonate, which is an additive used in the paper industry, thus substituting kaolin. It can be obtained by the following chemical reactions:

$$CaO + H_2O \rightarrow Ca(OH)_2$$
$$Ca(OH)_2 + CO_2 \rightarrow CaCO_3$$

Other possibilities are the injection of CO_2 near artificial reefs so that the growing of maritime algae is induced and, thus, promote the increase of fish population near those reefs. This is a process still under research and development.

Chapter 4

CONCLUSION

It was pointed out that separation processes to capture CO_2 and, thus, obtaining high purity, high concentrated CO_2 from fossil fuels require large amounts of energy. CO_2 capture and separation from large point source such as power plants can be achieved through continued research, development and demonstration, following the experiences and pilot projects described in section 2.2.

MEA has been, by far, the most commonly used absorbing solvent, and can be enhanced by additives to improve its performance. However, further research is still needed in order to obtain novel absorbing amines that can reduce the energy consumption in the solvent regeneration step and also reduce the corrosion characteristics of the solvents. In the near future, the efforts in R&D will possibly be centered on process optimization, while trying to demonstrate the concept at industrial scale in the range of 2-4 Mton CO_2/year. R&D in the long term will continue to be focused on obtaining newer solvents for absorption. The possibility of capturing carbon and fixing it as a solid or use it as a secondary product is considerably attractive.

Adsorption is, of course, an interesting technical option, but still needs that several technological barriers are overcome. The use of membranes brings considerable advantages but also needs that a combination of high gas fluxes and high selectivity are to be achieved simultaneously with low costs for these membranes. Nevertheless, very good perspectives exist concerning the use of this technology in the future. Also, it is considered that investment has to be made on research and development on alternatives for reutilization of CO_2, as geologic sequestration cannot be regarded as a stable long term solution.

REFERENCES

Abanades, J., Rubin, E., Anthony, E., 2004. Sorbent cost and performance in CO_2 capture systems. *Industrial and Engineering Chemistry Research*, 43(13): 3462–3466.

Abidi N, Sivadea A, Bourret D, Larbot A, Boutevin B, Guida- Pietrasanta F, Ratsimihety A, 2006. Surface modification of mesoporous membranes by fluoro-silane coupling reagent for CO_2 separation. *Journal of Membrane Science*, 270, 101–107.

Aboudheir, A., Tontiwachwuthikul, P., Chakma, A., Idem, R., 2003. Kinetics of the reactive absorption of carbon dioxide in high CO_2-loaded, concentrated aqueous monoethanolamine solutions. *Chemical Engineering Technology,* 58, 5195–5210.

Aitken, C., Koros, W., Paul, D., 1992. Effect of structural symmetry on gas transport properties of polysulfones. *Macromolecules*, 25(13), 3424–3434.

Anderson, J., Dixon, J., Maginn, E., Brennecke, J., 2006. Measurement of SO_2 solubility in ionic liquids, *Journal of Physical Chemistry*, B110, 15059-15962.

Anderson, J., Dixon, A., Brennecke, J., 2007, Solubility of CO_2, CH_4, C_2H_6, C_2H_4, O_2, and N_2 in 1-hexyl-3-methylpyridinium bis (trifluoromethylsulfonyl) imide: comparison to other ionic liquids, *Accounts Chemical Research,* 40, 1208-1216.

Anson, M., Marchese, J., Garis, E., Ochoa N, Pagliero C, 2004. ABS copolymer-activated carbon mixed matrix membranes for CO_2 / Ch_4 separation. *Journal of Membrane Science*, 243, 19–28.

Anthony, A., Maginn, E., Brennecke, J., 2002. Solubilities and thermodynamic properties of gases in the ionic liquid 1-n-butyl-3-

methylimidazolium hexafluorophosphate, *Journal Physical Chemistry, B* 106, 7315-7320.

Aoki, K., Kusakabe, K., Morooka, S., 2000. Separation of gases with an A-type zeolite membrane. *Industrial and Engineering Chemistry Research*, 39: 2245–2251.

Apichatachutapan, W., Moore, R., Mauritz, K., 1996. Asymmetric nafion/(zirconium oxide) hybrid membranes via inditu dol-gel vhemistry. *Journal of Applied Polymer Science*, 62(2), 417–426.

Bao L., Trachtenberg, M., 2005. Modelling CO_2-facilitated transport across a diethanolamine liquid membrane. *Chemical Engineering Science*, 60, 6868-6875.

Bates, E., Mayton, E., Ntai, I., Davis Jr., J., 2002. CO2 capture by a task-specific ionic liquid, *Journal American Chemical Society,* 124, 926-927.

Bernal M P, Piera E, Coronas J, Menendez M, Santamaria J, 2000. Mordenite and ZSM-5 hydrophilic tubular membranes for the separation of gas phase mixtures. *Catalysis Today*, 56, 221–227.

Bishnoi, S., Rochelle, G., 2002. Thermodynamics of piperazine/methyl diethanolamine/water/carbon dioxide. Industrial *Engineering Chemistry Research.* 41, 604–612.

Black, S., 2006. *Chilled ammonia scrubber for CO2 capture.* MIT Carbon Sequestration Forum VII. Cambridge, MA.

Boa, L., Trachtenberg, M., 2006. Facilitated transport of CO_2 across a liquid membrane: comparing enzyme, amine and alkaline, *Journal of Membrane Science*, 280, 330-334.

Cadena, C., Anthony, J., Shah, J., Morrow, T., 2004, Why is CO_2 so soluble in imidazolium-based ionic liquids?, *Journal American Chemical Society.* 126, 5300-5308.

CAPRICE, 2010. *CO2 Capture using Amine Processes International Cooperation and Exchange,* www.caprice-project.eu (accessed in April 2010).

Carapellucci, R., Milazzo, A., 2003. Membrane systems for CO_2 capture and their integration with gas turbine plants. *Proceedings of the Institution of Mechanical Engineers Part A: Journal of Power and Energy*, 217, 505–517.

Chaffee A L, 2005. Molecular modeling of HMS hybrid materials for CO2 adsorption. *Fuel Processing Technology*, 86(14-15), 1473–1486.

Chatti, R., Bansiwal, A., Thote, J., Kumar, V., Jadhav, P., Lokhande, S., Biniwale, R., Labhsetwar, N., Rayalu, S., 2009. Amine loaded zeolites for carbon dioxide capture: amine loading and adsorption studies, *Microporous and Mesoporous Materials,* 121, 84-89.

Chen, E., Rochelle, G., Seibert, F., 2006. *Pilot Plant for CO_2 Capture with Aqueous Piperzine/potassium Carbonate.* GHGT-8, Trondeim, Norway.

Chi, S., Rochelle, G., 2002. Oxidative degradation of monoethanolamine. *Industrial Engineering Chemistry Research*, 41, 4178–4186.

CO_2cesar (2010), CO_2cesar, 7th Framework Programme - *CO_2 enhanced separation and recovery, http://www.co2cesar.eu* (accessed in April 2010).

Damle, A., Dorchak, T., 2001. Recovery of carbon dioxide in advanced fossil energy conversion processes using a membrane reactor. *Journal of Energy and Environment Research*, 1(1), 77–89.

Davis, J., Rochelle, G., 2009. Thermal degradation of monoethanolamine at stripper conditions. *Energy Procedia,* 1 (1), 327–333.

Dong, J., Lin, Y., 1998. In situ synthesis of P-type zeolite-membranes on porous α-alumina supports. *Industrial and Engineering Chemistry Research*, 37, 2404–2409.

Druckenmiller, M., Maroto-Valer, M., 2005. Carbon sequestration using brine of adjusted pH to form mineral carbonates. *Fuel Processing Technology*, 86, 1599–1614.

Dugas, E., 2006. *Pilot plant study of carbon dioxide capture by aqueous monoethanolamine,* MSc Thesis, University of Texas at Austin, USA.

Edali, M., Aboudheir, A., Idem, R., 2009. Kinetics of carbon dioxide absorption into mixed aqueous solutions of MDEA and MEA using a laminar jet apparatus and a numerically solved 2D absorption rate/kinetics model. *International Journal Greenhouse Gas Control*, 3, 550–560.

Elwell, L., Grant, W., 2006. Technology options for capturing CO_2 -Special Reports. *Power,* 150 (8).

Energy Information Administration (EIA), 2006. *Annual Energy Outlook 2006.*

Essaki K, Nakagawa K, Kato M, Uemoto H, 2004. CO2 absorption by lithium silicate at room temperature. *Journal of Chemical Engineering of Japan*, 37(6), 772–777.

Fauth, D., Frommell, E., Hoffman, J., Reasbeck, R., Pennline, H., 2005. Eutectic salt promoted lithium zirconate: Novel high temperature sorbent for CO_2 capture. *Fuel Processing Technology*, 86(14-15), 1503–1521.

Feron, P., Hendriks, C., 2005. CO_2 capture process principles and costs. *Oil and Gas Science and Technology- Rev*, IFP 60(3): 451–459.

Figueroa J., Fout T., Plasynski S., McIlvried H., Srivastava R., 2008. Advances in CO_2 capture technology - The U.S. Department of Energy's Carbon Sequestration Program, *International Journal of Greenhouse Gas Control*, 2, 9–20.

Freguia, S, Rochelle, G., 2003. Modeling of CO_2 capture by aqueous monoethanolamine. *AIChE Symposium Series,* 49, 1676–1686.

Guha A., Majumdar S., Sirkar, K., 1990. Facilitated transport of CO_2 through an immobilized liquid membrane of aqueous diethanolamine. *Industrial Engineering Chemistry Research,* 29, 2093-2100.

Grande, C., Rodrigues, A., 2008. Electric swing adsorption for CO_2 removal from flue gases, *International Journal of Greenhouse Gas Control*, 2, 194-202.

Gray, M., Soong, Y., Champagne, K., Pennline, H., Baltrus, J., Stevens, R., Khatri, R., Chuang, S., Filburn, T., 2005. Improved immobilized carbon dioxide capture sorbents. *Fuel Processing Technology*, 86(14-15), 1449–1455.

Greer, T., Bedelbayev, A., Igreja, J., Gomes, J., Lie, B., 2010. A simulation study on the abatement of CO_2 emissions by deabsorption with monoethanolamine, *Environmental Technology*, 31(1), 107-115.

Gutowski, K., Maginn, E., 2008. Amine-functionalized task-specific ionic liquids: A mechanistic explanation for the dramatic increase in viscosity upon complexation with CO_2 from molecular simulation, *Journal American Chemical Society,* 130, 14690-14704.

Hakka, L., 2007. *Cansolv Technologies Inc.*

Haywood, H., Eyre. J., Scholes, H., 2001. Carbon dioxide sequestration as datable carbonate minerals–Environmental barriers. *Environmental Geology*, 41, 11–16.

Heintz, Y., Sehabiague, L., Morsi, B., Jones, K., Luebke, D., Pennline, H., 2009. Hydrogen sulfide and carbon dioxide removal from dry fuel gas streams using an ionic liquid as a physical solvent, *Energy Fuels,* 23, 4822-4830.

Hou, Y., Baltus, R., 2007. Experimental measurement of the solubility and diffusivity of CO_2 in room-temperature ionic liquids using a transient thin-liquid-film method, *Industrial Engineering Chemical Research,* 46, 8166-8175.

Huang, H., Shi, Y., Li, W., Chang, S., 2001. Dual alkali approaches for the capture and separation of CO_2. *Energy and Fuels*, 15(2), 263–268.

Huang, H., Yang, R., Chinn, D., Munson, C., 2003. Amine- grafted MCM-48 and silica xerogel as superior sorbents for acidic gas removal from natural gas. *Industrial and Engineering Chemistry Research*, 42(12), 2427–2433.

Hughes, R., Lu, D., Anthony, E., Macchi, A., 2005. Design, process simulation and construction of an atmospheric dual fluidized bed combustion system for *in situ* CO_2 capture using high-temperature sorbents. *Fuel Processing Technology*, 86(14-15), 1523–1531.

Husain, S., Koros, W., 2007. Mixed matrix hollow giber membranes made with modified HSSZ-13 zeolite in polyetherimide polymer matrix for gas separation. *Journal of Membrane Science*, 288, 195–207.

iCap, 2010. *iCap News and Highlights.* http://www.icapco2.org/news 1.html (accessed in October 2010).

Illing, G., Hellgardt, K., Wakeman, R., Jungbauer, A., 2001. Preparation and characterisation of polyaniline based membranes for gas separation. *Journal of Membrane Science*, 184, 69–78.

IPCC, 2005. Intergovernmental Panel on Climate Change (IPCC), *Special Report on Carbon Dioxide Capture and Storage.* Cambridge University Press, Cambridge, UK.

Kamps, A., Tuma, D., Xia, J., Maurer, G., 2003. Solubility of CO_2 in the ionic liquid [bmim][PF_6], *Journal Chemical Engineering Data,* 48, 746-749.

Kanel, J., 2003. *Overview: industrial application of ionic liquids for liquid extraction.* Presented at Chemical Industry Vision 2020, Technology Partnership Workshop, New York.

Kato, M., Nakagawa, K., Essaki, K., Maezawa, Y., Takeda, S., Kogo, R., Hagiwara, Y., 2005. Novel CO_2 absorbents using lithium-containing oxide. *International Journal of Applied Ceramic Technology*, 2(6), 467–475.

Kim, H., Lim, C., Hong, S., 2005. Gas permeation properties of organic-inorganic hybrid membranes prepared from hydroxyl-terminated polyether and 3-isocyanatopropyltriethoxysilane. *Journal of Sol-Gel Science and Technology*, 36(2), 213–221.

Kittel, J., Idem, R., Gelowitz, D., Tontiwachwuthikul, P., Parrain, G., Bonneau, A., 2009. Corrosion in MEA units for CO_2 capture: pilot plant studies. *Energy Procedia*, 1 (1), 791–797.

Knowles, G., Delaney, S., Chaffee, A., 2006. Diethylenetriamine[propyl (silyl)]-functionalized (DT) mesoporous silicas as CO_2 adsorbents. *Industrial and Engineering Chemistry Research,* 45(8), 2626–2633.

Knowles, G., Graham, J., Delaney, S., Chaffee, A., 2005. Aminopropyl-functionalized mesoporous silicas as CO_2 adsorbents. *Fuel Processing Technology*, 86, 1435–1448.

Knudsen, J., Jensen, J., Vilhelmsen, P., Biede, O., 2009. Experience with CO_2 capture from coal flue gas in pilot-scale: testing of different amine solvents. *Energy Procedia*, 1 (1), 783–790.

Kothandaraman A., 2010. *Carbon Dioxide Capture by Chemical Absorption: A Solvent Comparison Study*, Submitted to the Department of Chemical Engineering on May 20, 2010 in partial fulfillment of the requirements of the Degree of Doctor of Philosophy in Chemical Engineering Practice.

Kumełan, J., Tuma, D., Kamps, A., Maurer, G., 2010. Solubility of the single gases carbon dioxide and hydrogen in the ionic liquid [bmpy][Tf2N], *Journal Chemical Engineering Data,* 55, 165–172.

Kusakabe, K., Ichiki, K., Hayashi, J., Maeda, H., Morooka, S., 1996. Preparation and characterization of silica-polyimide composite membranes coated on porous tubes for CO_2 separation. *Journal of Membrane Science*, 115, 65–75.

Li, Y., Ding, M., Xu, J, 1997. Gas separation properties of aromatic polyetherimides from 1,4-bis(3,4- dicarboxyphenoxy)benzene dianhydride and 3,5-diaminobenzic acid or its esters. *Journal of Applied Polymer Science*, 63(1), 1–7.

Li, X., Hou, M., Zhang, Z., Han, B., Yang, G., Wang, X., Zou, L., 2008. Absorption of CO_2 by ionic liquid/polyethylene glycol mixture and the thermodynamic parameters, *Green Chemistry,* 10, 879-884.

Litynski, J., Klara, S., McIlvried, H., Srivastava, R., 2006. The United States Department of Energy's Regional Carbon Sequestration Partnerships Program: a collaborative approach to carbon management. *Environment International,* 32, 128–144.

Liu, N., Bond, G., Abel, A., McPherson, B., Stringer, J., 2005. Biomimetic sequestration of CO_2 in carbonate form: Role of produced waters and other brines. *Fuel Processing Technology*, 86, 1615–1625.

Luebke, D., Myers, C., Pennline, H., 2006. Hybrid membranes for selective carbon dioxide separation from fuel gas. *Energy and Fuels*, 20(5), 1906–1913.

Luis, P., Neves. L., Afonso, C., Coelhoso, I., Crespo, J., Garea, A., Irabien, A., 2009. Facilitated transport of CO_2 and SO2 through Supported Ionic Liquid Membranes. *Desalination*, 245, 485-493.

MacKenzie, A., Granatstein, D., Anthony, E., Abanades, J., 2007. Economics of CO_2 capture using the calcium cycle with a pressurized fluidized bed combustor. *Energy and Fuel*, 21, 920–926.

Mano, H., Kazama, S., Haraya, K., 2003. *Development of $_{CO2}$ separation membrane (1) Polymer membrane.* In: Gale J, Kaya Y, editors. Greenhouse gas control technologies. Elsevier Science, Ltd.

Manovic, V., Anthony, E., 2007. Steam reactivation of spent CaO-based sorbent for multiple CO_2 capture cycles. *Environmental Science and Technology*, 41, 1420–1425.

Maroto-Valera, M., Fauth, D., Kuchta, M., Zhang Y, Andresen, J., 2005. Activation of magnesium rich minerals as carbonation feedstock materials for CO_2 sequestration. *Fuel Processing Technology*, 86, 1627–1645.

Maroto-Valera, M., Tang, Z., Zhang, Y., 2005. CO_2 capture by activated and impregnated anthracites. *Fuel Processing Technology*, 86(14-15), 1487–1502.

NETL, 2008. *Carbon dioxide Capture by Absorption with Potassium Carbonate* (accessed April 2010).

Notz, R., Asprion, N., Clausen, I., Hasse, H., 2007. Selection and pilot plant tests of new absorbents for post-combustion carbon dioxide capture. *Chemical Engineering Research Design*, 85 (A4), 510–515.

NZEC, 2009. *The joint China–UK near zero emissions coal (NZEC) Initiative: Summary Report,* http://www.nzec.info/en/assets/Reports/China-UK-NZEC- English-031109.pdf (accessed October 2010).

Nomura, M., Yamaguchi, T., Nakao, S., 1997. Silicalite membranes modified by counterdiffusion CVD technique. *Industrial and Engineering Chemistry Research*, 36, 4217–4223.

Oyenekan, B., Rochelle, G., 2006. Energy performance of stripper configurations for CO_2 capture by aqueous amines. *Industrial and Engineering Chemistry Research*, 45(8), 2457–2464.

Nelson, M., 2004. *Carbon dioxide sequestration by mechanochemical carbonation of mineral silicates* - Final Report. USDOE.

Nelson, M., Prisbrey, K., 2004. *Mechanochemical methods for CO_2 sequestration.* SME annual meeting, Feb 23-25 2004, Denver, CO, United States. Preprints, 923–925.

Olajire, A., 2010. CO_2 capture and separation technologies for end-of-pipe applications – A review, *Energy*, 35, 2610-2628.

Okabe, K., Matsumija, N., Mano, H., Teramoto, M., 2003. *Development of CO_2 separation membranes (1) Facilitated transport membrane.* In: Gale J, Kaya Y, editors. Greenhouse gas control technologies. Elsevier Science Ltd.

Panahi, M., Skogestad, S., 2010. Economically Efficient Operation of CO_2 Capturing Process Part I: Self-optimizing Procedure for Selecting the Best Controlled Variables, *Chemical Engineering and Processing*, 2010.

Phan, A., Doonan, C., Uribe-Romo, F., Knobler, C., O'Keefe, M., Yaghi, O., 2010. Synthesis, structure and carbon dioxide properties of zeolitic imidazolate frameworks, *Accounts of Chemical Research*, 43, 58-67.

Pixton, M., Paul, D., 1995. Gas transport properties of polyarylates. Part I: Connector and pendant group effects. *Journal of Polymer Science Part B: Polymer Physics*, 33(7), 1135–1149.

Poshusta, J., Tuan, V., Pape, E., Noble, R., Falconer, J., 2000. Separation of light gas mixtures using SAPO-34 membranes. *AIChE Journal*, 46(4), 779–789.

Powell, Q., Qiao, G., 2006. Polymeric CO_2 /N_2 gas separation membranes for the capture of carbon dioxide from power plant flue gases. *Journal of Membrane Science*, 279, 1-49.

Rao, A., Rubin, E., 2009. A Technical, Economic, and Environmental Assessment of Amine-Based CO_2 Capture Technology for Power Plant Greenhouse Gas Control, *Environmental Science and Technology*, 43(3), 578-584.

Raeissi, S., Peters, C., 2009. Carbon dioxide solubility in the homologous 1-alkyl-3-methylimidazolium bis(trifluoromethyilsulfonyl)imide family, *Journal Chemical Engineering Data,* 54, 382-386.

Resnik, K., Yeh, J., Pennline, H., 2004. Aqua ammonia process for simultaneous removal of CO_2, SO_2 and NO_X. *International Journal Environmental Technology Management.* 4 (1/2), 89–104.

Rochelle, G., 2010. The Rochelle Lab, The University of Texas at Austin, *CO_2 capture,* http://www.che.utexas.edu/rochelle group/index.html (accessed in April 2010).

Rochelle, G., Chen, E., Dugas, R., Oyenakan, B., Seibert, F., 2006. *Solvent and process enhancements for* CO_2 *absorption/stripping.* In: 2005 Annual Conference on Capture and Sequestration, Alexandria, VA.

Rydén, M., Lyngfelt, A., 2006. Using steam reforming to produce hydrogen with carbon dioxide capture by chemical-looping combustion. *International Journal of Hydrogen Energy*, 31, 1271–1283.

Seibert, F., Chen, E., Perry, M., Briggs, M., Montgomery, R., Rochelle, G., 2011. UT/SRP CO_2 capture pilot plant – Operating experiences and procedures, GHGT-10. *Energy Procedia,* 4, 1616-1623.

Siriwardane, R., Shen, M., Fisher, E., 2003. Adsorption of CO_2, N_2 and O_2 on natural zeolites, *Energy & Fuels,* 17, 571-576

Song, C., 2006. Global challenges and strategies for control, conversion and utilization of CO_2 for sustainable development involving energy, catalysis, adsorption and chemical processing. *Catalysis Today*, 115, 2–32.

Shekhawat, D., Luebke, D., Pennline, H., 2003. A review of carbon dioxide selective membranes – A topical report. National Energy Technology Laboratory, United States Department of Energy.

Shin, E., Lee, B., 2008. High-pressure phase behavior of carbon dioxide with ionic liquids: 1-alkyl-3--ethylimidazolium trifluoro methanesulfonate, Journal Chemical Engineering Data, 53, 2728–2734.

Shen, K., Li, M., 1992. Solubility of carbon dioxide in aqueous mixtures of monoethanolamine with ethyldiethanolamine, *Journal Chemical Engineering Data,* 37, 96-100.

Singh, R., Way, J., McCarley, K., 2004. Development of a model surface flow membrane by modification of porous vycor glass with a fluorosilane. *Industrial and Engineering Chemistry Research*, 43(12), 3033–3040.

Sun, P., Grace, J., Lim, C., Anthony, E., 2007. Removal of CO_2 by Ca-based sorbents in the presence of SO_2. *Energy and Fuels*, 21, 163–170.

Stewart, C., Hessami, M., 2005. A study of methods of carbon dioxide capture and sequestration–the Sustainability of a photosynthetic bioreactor approach. *Energy Conversion and Management*, 46, 403–420.

Stolaroff, J., Lowry, G., Keith, D., 2005. Using CaO- and MgO-rich industrial waste streams for carbon sequestration. *Energy Conversion and Management*, 46. 687–699.

Tantekin-Ersolmaz, S., Atalay-Oral, Ç., Tatlier, M., Erdem-Senatalar, A., Schoemanb, B., Sterte, J., 2000. Effect of zeolite particle size on the performance of polymer-zeolite mixed matrix membranes. *Journal of Membrane Science*, 175, 285–288.

Trachtenberg, M., Tu, C., Landers, R., Wilson, R., McGregor, M., Laipis, P., Paterson, M., Silverman, D., Thomas, D., Smith, R., Rudolph, F., 1999. Carbon dioxide transport by proteic and facilitated transport membranes. *Life Support Biosphere Science*, 6, 293-302.

Uyanga, I., Idem, R., 2007. Studies of SO_2 - and O_2 -induced degradation of aqueous MEA during CO_2 capture from power plant flue gas streams. *Industrial Engineering Chemistry Research*, 46 (8), 2558–2566.

Vu, D., Koros, W., Miller, S., 2003. Effect of condensable impurity in CO_2 /CH4 gas feeds on performance of mixed matrix membranes using carbon molecular sieves. *Journal of Membrane Science*, 221, 233–239.

Vu, D., Koros, W., Miller, S., 2003. Mixed matrix membranes using carbon molecular sieves: II. Modeling permeation behavior. *Journal of Membrane Science*, 211, 335–348.

Xu, X., Song, C., Miller, B., Scaroni, A., 2005. Adsorption separation of carbon dioxide from flue gas of natural gas-fired boiler by a novel nanoporous "molecular basket" adsorbent. *Fuel Processing Technology*, 86(14-15), 1457–1472.

Wang, B., Cote, A., Furukawa, H., O'Keege, M., Yaghi, O., 2008. Collosal cages in zeolitic imidazolate frameworks as selective carbon dioxide reservoirs, *Nature*, 452, 207-212

Wang, M., Lawal, A., Stephenson, R., Sidders, J., Ramsshaw, C., 2011. Post-combustion CO_2 capture with chemical absorption: a state-of-the-art review, *Chemical Engineering Research Design*, 89, 1609-1621.

Willis, R., Benin, A., Low, J., Bedard, R., Lesch, D., 2006. *Annual Report, Project DE-FG26-04NT42121*, National Energy Technology Laboratory.

Wilson, M., Tontiwachwuthikul, P., Chakma, A., Idem, R., Veawab, A., Aroonwilas, A., Gelowitz, D., Barrie, J., Mariz, C., 2004. Test results from a CO_2 extraction pilot plant at boundary dam coal-fired power station. Energy, 29, 1259–1267.

Yamasaki, A., 2003. An overview of CO_2 mitigation options for global warming – Emphasizing CO_2 sequestration options. *Journal of Chemical Engineering of Japan*, 36(4), 361–375.

Yang, W., Ciferno, J., 2006. Assessment of Carbozyme enzyme-based membrane technology for CO_2 capture from flue gas. DOE/NETL 401/072606.

Yang, H., Xu, Z., Fan, M., Gupta, R., Slimane, R., Bland, A., Wright, I., 2008. Progress in carbon dioxide separation and capture: a review. *Journal of Environmental Studies*, 20, 14-27.

Yeh, J., Resnik, K., Rygle, K., Pennline, H., 2005. Semi-batch absorption and regeneration studies for CO2 capture by aqueous ammonia. *Fuel Processing Technology,* 86 (14–15), 1533–1546.

Yoshitake, H., Yokoi, T., Tatsumi, T., 2003. Adsorption of chromate and arsenate by amino-functionalized MCM-41 and SBA-1. *Chemistry of Materials*, 15, 4536–4538.

Ziaii, S., Rochelle, G., Edgar, T., 2009. Dynamic modelling to minimise the energy use for CO_2 capture in power plant aqueous monoethanolamine. *Industrial Engineering Chemistry Research*, 48, 6105–6111.

Zimmerman, C., Koros, W., 1999. Entropic selectivity analysis of a series of polypyrrolones for gas separation membranes. *Macromolecules*, 32(10), 3341-3346.

INDEX

C

D

E

F

G

N

O

P

R

S

T

U

V

W

Z